WITHDRAWN

J
599
BAN

D1307334

Bighorn sheep

BIGHORN SHEEP

BY BRETT BANNOR

**Endangered
Animals &
Habitats**

LUCENT
BOOKS ®

THOMSON
—✦—™
GALE

San Diego • Detroit • New York • San Francisco • Cleveland • New Haven, Conn. • Waterville, Maine • London • Munich

On cover: A Rocky Mountain bighorn
sheep surveys his domain.

For more information, contact
Lucent Books
27500 Drake Rd.
Farmington Hills, MI 48331-3535
Or you can visit our Internet site at http://www.gale.com

LIBRARY OF CONGRESS CATALOGING-IN-PUBLICATION DATA

Bannor, Brett.
    Bighorn sheep / by Brett Bannor.
        p. cm. — (Endangered animals and habitats)
Summary: Describes the physical characteristics and habits of bighorn sheep, threats
to their health and habitats, and efforts to protect these wild mammals.
    Includes bibliographical references and index (p.   ).
    ISBN: 1-56006-887-6 (hardback: alk. paper)
    1. Bighorn sheep—Juvenile literatue. [1. Bighorn sheep. 2. Sheep. 3. Endangered
species.] I. Title. II. Endangered animals & habitats
    QL737.U53 B34 2003
    599.649'7—dc21

                                                                    2002004884

Printed in the United States of America

Contents

Introduction

WHEN THE PIONEERS and settlers opened up the American West in the nineteenth century, their foremost interest was survival, not conservation. Because of unregulated hunting and human intrusions into the bighorn's habitat, America's wild sheep began to vanish from several of the western states where they had formerly ranged. By the mid-1930s, bighorn were extinct in Washington state, and all were gone from west Texas by about 1960.

Fortunately, the bighorn's distribution was so vast that they did not disappear everywhere. These animals continued to thrive in the core of their range—the Rocky Mountain states such as Colorado and Wyoming—after they had vanished from other places. The bighorn benefited when a conservation ethic developed in the early twentieth century that led to a ban on uncontrolled bighorn hunting.

Protection of unspoiled western habitats, which are home to a variety of wildlife besides the bighorn, also became a priority. In 1872 the nation's first national park, Yellowstone National Park, was established in northwestern Wyoming and parts of adjacent Idaho and Montana. The park lies in the center of the bighorn's range. By 1915, ten more national parks, monuments, and wildlife refuges had been created within the region inhabited by bighorn. More ecologically rich lands continued to be added to the United States' system of parks and refuges, and by 1972, there were thirty-one federally protected areas within the bighorn's historic U.S. range, not to mention the numerous state parks and national forests that had also been founded.

Canada and Mexico likewise designated sanctuaries to protect the bighorn.

Yet, even with hunting regulations and the establishment of refuges, bighorn sheep continued to decline in numbers. In 1980, the Colorado Division of Wildlife issued a special report revealing that, despite protection and management practices intended to assist the bighorn, the number of these animals in the state had steadily dropped over the previous century. Furthermore, the report noted that only recently had research begun to answer basic questions about the bighorn's life.

Armed with new information, scientists began to understand how and why disease played a significant role in keeping the bighorn population low. They learned why translocation attempts—instances in which bighorn were moved from one place to another for the purpose of starting new herds—often failed. And they began to realize

A bighorn ram displays his lithe and muscular body. Bighorn populations have declined, despite conservation efforts.

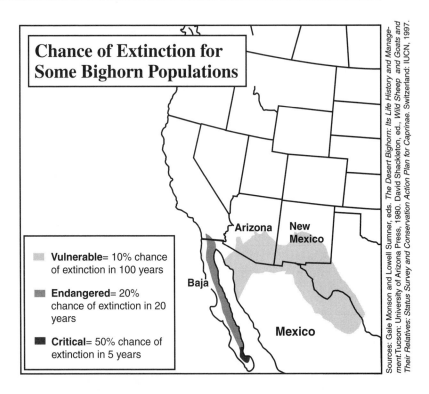

Chance of Extinction for Some Bighorn Populations

Vulnerable= 10% chance of extinction in 100 years

Endangered= 20% chance of extinction in 20 years

Critical= 50% chance of extinction in 5 years

Arizona New Mexico

Baja

Mexico

Sources: Gale Monson and Lowell Sumner, eds. *The Desert Bighorn: Its Life History and Management.* Tucson: University of Arizona Press, 1980. David Shackleton, ed., *Wild Sheep and Goats and Their Relatives: Status Survey and Conservation Action Plan for Caprinae.* Switzerland: IUCN, 1997.

what new courses of action needed to be taken to ensure the bighorn's survival.

More than two decades after the Colorado report came out, field studies and projects involving the bighorn continue. Yet in Colorado, and in other parts of the bighorn's range, challenges remain and many issues must be addressed before America's wild sheep are safe. Not all the proposals for managing bighorn are without controversy. For example, in places where small herds are preyed on by mountain lions, some wildlife biologists have suggested that the big cats need to be culled, that is, some of them killed, so that the bighorn may prosper. However, some activists believe that it is wrong to eliminate one type of animal to preserve another.

Another complication in bighorn management is that not all bighorn are subject to the same threats because of their broad range, so challenges to protecting these animals may not be the same in all regions. Nor is the bighorn even considered to be endangered everywhere; its status varies from place to place. The World Conservation Union (IUCN) divides the term *endangered* into designations based on specific criteria derived from mathematical probability models. A population of organisms is designated as critical if there is a 50 percent chance it will become extinct within five years. The Weems' bighorn of the southern part of the Baja Peninsula in Mexico is listed as critical; extinction may be imminent. The IUCN category *endangered* means there is a 20 percent chance that the population will go extinct within twenty years, and the peninsular bighorn of southern California and northern Mexico has this ranking. Finally, the designation *vulnerable* means that there is a 10 percent chance that a population will no longer exist within one hundred years. The desert bighorn of Arizona, New Mexico, and the adjacent Mexican states have been classified as vulnerable. Everywhere else, bighorn populations meet the IUCN's category of *lower risk,* which means that while numbers may be declining, as in Colorado, the chance of extinction is not yet high.

Thus, although the bighorn is one species of animal, it shows all possible degrees of scarcity, depending on what state or region it lives in. For scientists, then, it is important to understand the natural history of bighorn everywhere they live, not only in one or two refuges. Understanding the biology of these wild sheep is also necessary so that conservationists are aware of the threats bighorn face in different places and can plot sound strategies for protecting these beautiful animals.

1

Bighorn Natural History

FROM THE WESTERN Canadian provinces of British Columbia and Alberta, down through fifteen western states in the United States, and south into the Baja and Sonoran regions of Mexico lives a species of wild sheep known as the bighorn sheep (also called the bighorn). Its scientific name is *Ovis canadensis*. Ovis is a Latin word meaning "sheep," while canadensis simply means "of Canada," even though bighorn also live in the United States and Mexico.

The wild sheep living in Mexico, Texas, southern New Mexico, Arizona, southern California, southern Nevada, and Utah are called "desert bighorn sheep" by zoologists. Those that live in Canada, Washington, Oregon, Idaho, Montana, Wyoming, Colorado, and northern California are termed northern bighorn or mountain bighorn sheep.

Because they have such different habitats, desert bighorn and mountain bighorn are different in many ways. Studies of the ecology of the bighorn's habitat in both alpine and arid regions have enabled scientists to develop a more complete picture of the challenges these animals face every day in their natural habitats.

Size and appearance

The size and appearance of bighorn sheep vary depending on the habitat they live in. However, all bighorn are considered to be large animals. The grandest males, called rams, measure nearly six-and-one-half feet long and stand over three-and-one-half feet high at the shoulder. Female bighorn, also called ewes, tend to be smaller. The largest ewes are just over six feet long and stand three feet high at the shoulder.

The bighorn of the southwestern deserts weigh less than their cousins of the northern Rockies. While a desert ram is about 150 pounds, a typical mountain male weighs approximately 175 pounds. A typical desert ewe is about 115 pounds, while the upland ewes weigh around 130 pounds. The largest individual ever recorded was a 320-pound mountain ram.

Differences in the size of bighorns based on where they live is an example of Bergmann's Rule: In most mammals

Source: Gale Monson and Lowell Sumner, eds. *The Desert Bighorn: Its Life History and Management.*Tucson: University of Arizona Press, 1980.

How Many Subspecies?

There is wide agreement among zoologists that bighorn should be considered as belonging to one of two broad ecological types: the mountain bighorn and the desert bighorn. There is a great deal of disagreement, however, concerning the further division of bighorn into subspecies. For example, the now extinct populations of bighorn that originally lived in the western Dakotas and eastern Montana were often called Audubon's bighorn. In Wyoming, Colorado, Idaho, and western Montana, the animals are sometimes called Rocky Mountain bighorn. Farther west, in British Columbia and Washington, is the northern portion of the range of the California bighorn. Each of these three were once considered distinct subspecies.

Recently, scientists John Wehausen and Rob Roy Ramey carefully examined the skulls and horns of nearly seven hundred museum specimens of bighorn. They concluded that Audubon's bighorn, Rocky Mountain bighorn, and bighorn from Washington and Canada were not three subspecies, but only one. Since there is still disagreement on this matter, and since further research may lead to new conclusions, it seems best to use the terms "mountain bighorn" or "desert bighorn" and leave finer distinctions for specialists.

and birds, populations that live in colder climates are larger than those from warmer regions. According to Bergmann's Rule, because a larger body loses proportionately less heat than a smaller one, an individual from the cold northern climate benefits by being bulkier and retaining more of its body heat. On the other hand, in a hot, dry desert, it is advantageous to be smaller so that body heat is more quickly dispersed. Thus, bighorn of alpine Canada outweigh those in arid Mexico.

A bighorn's fur is brown with a light trim of creamy white patches on the rump, the muzzle, and the back of the legs. Generally desert sheep are somewhat lighter than their mountain cousins. This is true for many species of animals; individuals living in dry deserts tend to be lighter colored than those in more humid habitats such as forests. Scientists call this Gloger's Rule, and they think that the differences in color have to do with camouflage. A lighter color helps desert bighorns to blend in with their desert surroundings, thereby reducing their vulnerability to predators.

About those horns

The name *bighorn sheep* is somewhat misleading. Although females have horns, only adult males possess the "big" horns—massive, spiral structures that coil backward and then forward again. In a large adult Rocky Mountain bighorn ram, the horn length is over three-and-a-half feet. A large set of horns may weigh nearly forty-five pounds.

Because desert bighorn tend to be smaller than their northern cousins, desert rams usually have slightly smaller horns. Among the biggest desert males, average horns are less than forty-one inches long.

Although a horn appears to be in one solid piece, it is actually a two-part structure, consisting of a short, bony core attached to the skull and a sheath of keratin covering this

The size and shape of the horns make it easy to identify the male of the bighorn species.

core. Much of the bighorn's horn is actually made up of this sheath. Horns grow annually, but once a ram has reached seven or eight years of age, yearly horn growth is minimal from that time on. It appears that horns respond to food intake; the higher the consumption of quality food, the faster they grow.

Horns may be broken, as sometimes happens to rams in combat, but they are never shed in the manner deer annually drop their antlers. Often only the tips of horns break off, in which case they are said to be "broomed." A broken horn will continue to grow, but it will not attain the same size and symmetry of an unbroken horn.

The horns atop the heads of mature females are also curved, but measure under a foot in length. Desert ewes have slightly more curved horns than those of mountain females.

Bighorn habitats

Bighorn typically inhabit rugged terrain—such as alpine meadows, grassy slopes, and foothills—which is why the northern populations are often called "mountain sheep." Mountain bighorns are found at elevations ranging from 1,500 feet to 10,800 feet above sea level.

In winter, mountain bighorn spend over 85 percent of their time within about 330 feet of canyons, cliffs, and other steep places. These inclined features are called escape terrain, because bighorn can easily bound up and down these slopes to escape predators that cannot negotiate the rugged ground as easily. In summer, lack of snow makes it easier to run away from predators on level ground, so during the warmer months mountain sheep may stray as far as half a mile from escape terrain.

The mountain bighorn are not the only sheep that live among sharp grades and precipices. The desert bighorn frequents the deep canyons, sheer cliffs, and rock outcroppings of the isolated mountain ranges of the Southwest, such as the Kofa Mountains of Arizona, the San Andres Mountains of New Mexico, the Little San Bernardino Mountains of California, and the appropriately named Sheep Range of

Nevada. Desert bighorns, in fact, are found both high and low: Some live in Death Valley, California, 256 feet below sea level, while others inhabit California's White Mountains at elevations as high as fourteen thousand feet.

Zoologists have been intrigued by exactly how steep a landscape bighorn may inhabit. In one part of California, for example, researchers reported that bighorn sheep activities were not impaired by slopes of up to 80 percent. (An 80 percent slope means that for every ten feet of land measured horizontally there is a drop of eight feet in elevation.) Bighorn can easily traverse grades twice as steep as inclines used by deer and four times as steep as those used by domestic cattle.

Bighorn sheep are equally agile in other areas of the United States. In Colorado, for example, sheep commonly

The Bighorn and Its Relatives

Bighorn—like other sheep and goats, plus camels, giraffes, deer, hippopotamuses, pigs, buffalo, and antelope—are mammals and members of the order Artiodactyla. This is a Greek word that means "having an even number of hooves."

The order Artiodactyla is divided into several families. The deer, for example, are in a family called Cervidae, while the giraffes are members of family Giraffidae. The largest family of Artiodactyla is Bovidae; it includes sheep, goats, antelope, and cattle. It is the cattle that give the family its name, which means "cow" in Latin.

While there might appear to be great differences between a bighorn, a gazelle, and domestic cattle, all are placed in the family Bovidae because they have horns consisting of two parts: a bony core and a horny sheath. In some Bovids—like the bighorn sheep—both sexes have horns; in others only the males are adorned.

Besides the bighorn, there are six other species of sheep in the genus *Ovis*. One of these relatives is the domestic sheep, *Ovis aries*. The name *aries* is also Latin; it means "a ram." Domestic sheep are found nearly worldwide, having been transported to all parts of the earth by people.

Most wild sheep are from mountainous regions of Eurasia. The only other North American species besides the bighorn is the thinhorn sheep (*Ovis dalli*). It is native to Alaskan mountain ranges such as the Wrangells, the Kenai, and the Brooks range, and also is found in northwestern Canadian uplands such as the Mackenzie Mountains and the MacArthur Ranges.

Three bighorn lambs maneuver surefootedly across rocky terrain.

inhabit areas with slopes of 61 percent to 80 percent but avoid areas with slopes of less than 20 percent.

While the narrow cloven hooves of a bighorn provide traction and grip as the animal travels over treacherous terrain, the bighorn possesses another important asset in dealing with rugged habitat—its memory. By learning the location of all the crevices and crannies found where it lives, a bighorn is able to quickly negotiate up, down, and around all the obstacles in its homeland. A running bighorn does not slow down or stop to check its footing, and these animals have a high degree of fidelity to their ancestral home ranges. That is, the offspring tend to stay in the same area where their parents live, and thus knowledge of landscape features is passed along from generation to generation.

The area of land upon which an animal conducts its daily activities is called its home range. There is little information on home range size for mountain sheep, but in Utah, scientists monitored desert bighorn fitted with radio collars and learned that males have an average home range of twenty-three-and-a-half square miles. The ewes' home ranges were much smaller, with an average size of just over nine square miles. This is consistent with what scientists have learned about many other species of mammals in which the males and females do not associate regularly; often male home ranges are larger. Zoologists theorize that males benefit by having as many different females as possible living within their haunts, since this provides them the greatest number of potential mates.

It is also thought that the more scattered the resources are, the larger the area the bighorn will need to roam to meet all its survival needs. These resources include shelter, food, and water. Food and water may be especially difficult for bighorn to procure at certain times of year, and they may have to travel farther to find it.

The harsh seasons

Bighorn from mountains and those from deserts both must deal with climates that can be harsh at certain times of the year. The type of hardship faced depends on where a particular population lives.

Mountain bighorn must cope with an environment where winters are cold, with overnight low temperatures frequently under 20°F (−6.6°C). Snowfall may come at any time from early fall to late spring, and it may be severe. Several of the record snow accumulations for the United States took place in western mountain regions inhabited by bighorn. For instance, the most snow to drop in one year was at Mt. Rainier, Washington, where over ninety-three feet of snow fell in the 1971–1972 season. Similarly, the record for most snow in twenty-four hours is six feet four inches in the foothills of the Colorado Rockies in 1971. Deep snow makes it difficult for bighorn to find food, so much so that they generally avoid areas where the snow depth is a foot or more.

Thus, mountain sheep must find snow-free areas in the alpine winters. There are two methods by which they do this. One way to escape the deep snow is altitudinal migration (moving uphill or downhill). Each fall, bighorn are able to escape some of the rigors of heavy snow by simply moving downhill until they reach areas where the air is slightly warmer, snow is shallow or absent, and food is available. In the spring, when highland snows recede, the animals move back uphill to take advantage of the fresh growth of alpine vegetation.

Besides altitudinal migration, bighorn may also simply move around the hill. In winter it is much more common to find mountain sheep on steep slopes that face south, southeast, or southwest than on northerly facing inclines. By favoring slopes that face the winter sun, the sheep are taking advantage of a microclimate that is less likely to have standing deep snow inhibiting feeding. In spring, food plants will begin to grow on these southerly slopes sooner than on the mountain sides that face away from the sun.

The severest weather challenges facing desert bighorn are heat and dry conditions. Sheep in the Mojave Desert, for example, inhabit a region in which midday temperatures in the summer typically top 100°F (37.7°C) and an-

A mountain bighorn male makes his way through knee-deep snow.

nual rainfall is less than an inch. Because desert bighorn inhabit a climate that is very arid, the lack of water has the most impact on a bighorn population's survival.

In the summer, some waterholes may be dry, and those that do contain water may be located far apart. Because of the scarcity of water in the summer, desert bighorn may need to go five to fifteen days without drinking. One of the ways desert sheep compensate for limited water resources is by drinking heavily on those rare occasions when water is available. One study found that when desert bighorn go without water for five days, they lose about one-fifth of their body weight, or as much as thirty pounds for an adult ram. Within an hour after drinking on the sixth day, however, they had gained back all their original weight.

In some cases, humans have had a positive influence on the availability of water for desert bighorn. Reservoir and stock ponds intended to supply water needs for humans and their livestock also provide drinking needs for some wild sheep.

When water is scarce or absent, desert bighorn can conserve body fluids by reducing activity in the heat of the day. The weather does not, however, need to be very hot before they refrain from exertion. In Arizona it was found that bighorn bedded down in the shade for more than seven hours a day when temperatures rose above the relatively cool mark of 66°F (18.8°C).

Another desert survival strategy is to obtain moisture from eating plants. For instance, in Arizona, bighorn are known to consume barrel cacti in the driest months to obtain the precious fluids stored by this plant.

When studying the mountain bighorn, some scientists reported never seeing them drink. It is assumed that these animals are able to satisfy their water requirements in winter by consuming snow and ice. In the summer, they acquire the fluids they need from the plants they eat.

Even if bighorn find plenty of food and water and survive the harsh seasons, they typically only live ten years. However, in unusual cases rams have been known to live twenty years and ewes twenty-four years.

Food

Bighorn sheep are herbivorous, meaning they feed on vegetation. In the Sonoran Desert of Arizona, scientists identified 121 different species of plants that the desert sheep ate. Grasses were only a small part of the diet. *Forbs,* which are herbs and other small broadleaf plants, were more commonly eaten. Two-thirds of the bighorn meal was vegetation called *browse,* which is the leaves, shoots, and twigs of shrubs and other woody plants. The favorite food plants of these desert bighorn include propeller bush, desert mallow, and white ratany.

The diet is different for sheep in the mountains. Zoologists in British Columbia counted seventy-nine species of plants eaten by the mountain sheep they studied. Not only is this far fewer plants than were recorded consumed by the Arizona bighorns but also the relative preferences of browse and grass were reversed. These mountain animals ate less browse and more grass than their desert kin. In fact, one particular species of grass, the bluebunch wheatgrass, provided over one-fifth of the diet for these Canadian bighorn.

A Four-Chambered Stomach

Whether dining on grass or browse, a diet of rough vegetation provides challenges to the bighorn's digestive system. Like goats, cattle, antelope, deer, and giraffes, sheep are ruminants; they have a four-chambered stomach and a complex system for digesting their food. While eating, a bighorn will first quickly bite off and chew leaves, grass, twigs, and other plant parts. These bits of vegetation are swallowed and passed down to the first chamber of the stomach, called the rumen. After eating, when the animal lies down to rest, it regurgitates the food back into its mouth, where it is chewed more thoroughly. This is called chewing the cud. After this second round of chewing, food passes down into the second, third, and fourth chambers of the stomach for further processing before digestion is complete.

This long procedure for bighorn and other ruminants to process their food is necessary because most plants contain large amounts of chemically complex substances that are difficult to digest. The bighorn rumen alleviates this problem, because it is loaded with bacteria and microscopic, single-celled animals that have the ability to break down those troublesome compounds.

Whether feeding or performing any of its other daily activities, a bighorn is likely to be part of a flock of individuals doing the same thing. The bighorn is a social animal.

The social life

Zoologists have long wondered why so many species of animals live in groups. Two common theories are that animals are social either because this increases their success at finding quality food or because the more individuals there are in a group, the better chance there is that a predator will be spotted before it can make a kill. In the case of the bighorn, reduced risk of predation is probably most responsible for its gregarious habits.

Another advantage to group living is that social animals may groom each other. For bighorn sheep, one individual may use its tongue to groom the face, ears, or horns of another; these are spots the groomed animal cannot reach itself. This is called *allogrooming,* and it is thought to be a means of eliminating ticks or other small invertebrates that cling to the bighorn's body.

For much of the year, rams and ewes do not socialize with members of the opposite gender. The bighorn thus forms what scientists call sexually segregated groups. Ewes, along with lambs and immature rams, form one type of flock, called a nursery group. The other type of flock, which consists of adult rams, is called a bachelor group.

In an attempt to figure out why sexual segregation occurs, scientist Kathreen Ruckstuhl observed both female and male groups as they walked, grazed, or reclined. She noticed that rams spent more time resting than did ewes, and that ewes spent more time walking and feeding than the males. It may be that bighorn form separate male and female groups simply because the difference in movement patterns makes it difficult for the sexes to stay together.

A small difference has been noted in group size between mountain bighorn and desert bighorn. Bachelor mountain sheep flocks in Yellowstone National Park, Wyoming, had an average of about five and a half members, with nursery groups containing an average of eight individuals. For

Partners

Bacteria and protozoans—one-celled animals—often live inside a larger animal. In some cases, this causes the infested animal to sicken or even die because its vital organs or immune system cannot function with the added strain caused by the tiny invaders. This is called parasitism, and the small organisms within the larger one are termed parasites.

The inhabitants of a bighorn's rumen, however, are not parasites. Rather than harming the bighorn, these bacteria and protozoans break down the tough cell walls of the plants the bighorn eats. This allows the sheep to digest the food it needs to fuel its daily activities. In turn, the bacteria and protozoans benefit from the warmth and security the bighorn's rumen provides. Such a relationship, in which both parties benefit from the arrangement and neither is harmed, is called mutualism.

desert bighorn in Nevada, bachelor groups averaged four members and nursery groups nearly six animals. The larger group sizes for mountain herds might be further evidence that bighorn are social as a defense against predators. Since the mountain bighorn's habitat often contains more predators, they would benefit from having more watchful eyes in the group than would desert bighorn herds.

The mating season

The bighorn have an intricate mating system. First, bighorns have both a rut and a prerut (pronounced "pre-rut.") A rut, or rutting season, is the time when males and females actually mate. One to two months before mingling with the females, the males also participate in prerut, during which the bachelor groups congregate and have quarrels and contests to determine dominance hierarchies. Simply put, that means they are challenging each other to see which ram will earn the highest ranking, which is second, and so on down the line.

It is during this prerut, not during the mating season itself, that bighorn rams engage in spectacular head clashes. When two big males slam headfirst into each other, the sound can often be heard over a mile away. Although the clashes appear painful, usually injuries result only if the contesting rams fail to strike each other horn to horn. If one set of horns meets the other slightly askew, one or both males may be severely pummeled in the face.

Typically, one ram will only challenge another if they are fairly evenly matched in horn size. Thus, a two-year-old will not pick a fight against a seven-year-old with much more formidable headgear. Although there are contests between males close together in age, overall the oldest males with the largest horns are the most dominant animals.

When the males finally congregate with the ewes for the actual rut, big rams commonly engage in tending. This is the behavior in which a ram stays close to a female in heat and keeps other males away from her. Alternatively, a ram might try blocking, by actually prodding a ewe to move her away from the spot where the other males are seeking mates. A single ram may tend or block one or several ewes for up to three days.

By the time a male is two years old, it is sexually mature but not socially mature. This means that the young male is prevented from acquiring a more thorough tending or blocking because of the dominance and aggression of big rams that are six years of age or older.

There is, however, a third mating tactic open to these low-ranking males; it is called *coursing*. In coursing, a

Two mountain bighorn rams charge each other in a prerut contest of dominance.

A mountain bighorn ram prepares to mate with a ewe during the late fall breeding season.

young ram will vigorously fight an older one to earn the few seconds it needs to quickly mate with whatever ewe it can gain access to. These rams are surprisingly successful; one study concluded that 44 percent of all lambs are fathered by coursing rams. This is a large number, considering that the most dominant ram in the rut, which does not need to resort to coursing, may by himself be responsible for over 40 percent of the actual matings with ewes.

There is a notable cost to coursing, however. Two zoologists who monitored bighorn mating wrote: "Injury from falls and horn blows during coursing brawls may cause death, handicap future mating competition, or increase the risk of predation."[1] An increased risk of predation occurs because a ram badly injured in a brawl may be unable to run away from a mountain lion or other predator.

On the one hand, a young ram may need to resort to coursing if he wishes to mate; on the other hand, doing so might cost him his life. Commenting on this trade-off, scientists call the strategy of coursing "high-gain and high-risk."[2]

The life of the lamb

Mountain bighorn have a short breeding season that occurs in November and December. Desert bighorn, in contrast, may mate anytime from July through December. The varying dates when lambs are born reflects this difference. In California's Mojave Desert, for example, most births take place in February and March, while in the Rocky Mountains, mid-June is the peak of lambing season.

The shorter lambing season for mountain bighorn is probably due to the dietary needs of nursing ewes. A mother sheep needs to consume plenty of vegetation while she is nursing her offspring. Mountain sheep thus have their lambs in late spring, when nourishing new plant growth appears in the highlands.

A bighorn ewe generally first gives birth when she is two or three years old, and in exceptional cases she may continue to bear young until she is thirteen years old. The gestation period is 170 to 180 days, or about six months. Usually there is only one offspring; twins are rare.

Bighorn ewes usually isolate themselves from the rest of the flock just prior to giving birth. One reason they are thought to do this is that if the ewe is by herself, there is less chance she will be noticed by a predator while she and

 Lost Lambs

While there are many differences between mountain and desert bighorn, it would be an oversimplification to suggest that all differences in bighorn populations are based on which habitat the sheep live in. The bighorn in Montana and in northern New Mexico are both from the mountain race, but zoologist Christine Hass, who compared the nursing behavior of lambs in the two regions, found an odd contrast. In the Montana herd, it was common for a lamb to be nursed by a ewe other than the lamb's mother. In fact, some of the lambs suckled as much as 85 percent of the time from what Wass called an "alien female," that is, a ewe that did not give birth to the lamb she permitted to nurse.

In New Mexico, however, no lamb was seen nursing at the side of any ewe *other* than its own mother. It was suggested that the Montana ewes were receptive to nursing lambs other than their own because there was higher loss of lambs to predation than was the case with the New Mexico animals. A lactating ewe that had lost its own offspring might simply be extending its maternal instinct by nursing another lamb that had not perished.

her newborn offspring are particularly vulnerable. It is also possible that being alone together strengthens the bond between mother and young.

At birth, the lamb weighs six to ten pounds. Baby bighorn are precocious and rise to their feet within hours of birth. In fact, during its first day of life, the lamb is par-

The gestation period for bighorn sheep is about six months. Here a two-week-old bighorn lamb nuzzles its mother.

ticularly active and playful, as it develops the strong limbs and reflexes it will need to follow its mother through rough country.

When mother and baby rejoin the flock, the lamb continues its playful ways. One study showed that male lambs played more than females did, and that a lamb was most likely to choose as its playmate the youngster in the flock closest to itself in age. Some zoologists believe that by playing, lambs are developing the skills and motor coordination they will need to have as adults.

Although bighorn are known to die from falls suffered in their rocky homes, there are few eyewitness accounts of fatal accidents. It is likely that lambs, because of their inexperience and their growing bodies, suffer a disproportionate number of these falls. It was a lamb that was killed in perhaps the most detailed note of a bighorn fall reported by scientist Gary Brundige. While conducting observations in the Black Hills of South Dakota, Brundige witnessed a two-week-old lamb tumble to its death as the youngster attempted to follow its mother and lamb up the rim of a nearly five-hundred-foot-deep canyon. The lamb first fell about 130 feet; it was injured but still alive. Its mother quickly moved down to where the lamb had fallen into some bushes. As the lamb thrashed about, it fell another 130 feet. After this fall the lamb did not stir, although once again its mother rushed to its side.

The scientist gathered the body for a necropsy, an examination of a deceased animal by a veterinarian. The vet found that the lamb had suffered a fractured jaw, a bruised cranium, multiple fractures of the liver, and extensive internal hemorrhaging.

The occasional fatal fall by a bighorn lamb is a case of natural mortality. All living things eventually perish. What concerns scientists and conservationists today is that in some places bighorn are vanishing due to an unnatural cause, namely the impact of human beings on the bighorn population.

2

Hunting and Poaching of Bighorn

BIGHORN SHEEP HAVE been hunted by humans for thousands of years. At one time, people hunted bighorn for food, but by the twentieth century most hunting was for sport. Today some shooting of bighorn is legal and regulated by state governments. However, the illegal killing of bighorn poses a threat to some of the bighorn populations.

Early evidence of bighorn

About 2 million years ago, ancestors of the bighorn crossed into North America from northeastern Asia. For most of that time, bighorn, like all other animals in the New World, lived in a land uninhabited by people. This isolation of America's wildlife from human contact ended between fourteen and thirty-five thousand years ago. At that time, tribes of hunter-gatherers migrated into America from Asia, just as the bighorn had done much earlier.

Proof that at least some of North America's aboriginal tribes hunted bighorn is provided by the abundance of rock drawings made by the continent's earliest humans, especially in the canyons and caves of the Rocky Mountain region. Either by painting or by chipping away outer layers of stone, these artists would create images on a rock surface. There are more drawings of bighorns than of any other type of animal, and often the wild sheep are depicted as quarry of the hunt. The earliest drawings show the humans hurling spears or atlatls (spear-headed throwing sticks) at the bighorn. About two thousand years ago, Native Americans invented bows and arrows. Corresponding to this development, the rock images of this era depict bighorn being pursued by bow hunters.

Bighorn sightings

The first Europeans to observe bighorn were Spanish explorers and missionaries who came to California and other parts of the Southwest in the sixteenth century. It was not until 1702, however, that a missionary named Father Francisco Maria Piccolo published an account of the bighorn. Later still, in 1757, the first book with an illustration of one of these animals appeared. This volume's title, translated into English, was *A Natural and Civil History of California,* and it referred to the bighorn as the "California deer."

In 1803, parts of a deceased bighorn were sent to the British Museum in London. This was the first time museum curators and zoologists had the opportunity to inspect North American wild sheep. Zoologist George Shaw was

The abundant number of rock drawings of bighorn sheep reveals the importance of the animal to Native American Indians.

Bighorn Before Columbus

Several groups of aboriginal North Americans are known to have hunted bighorn. In Alberta, anthropologists unearthed the remains of three bighorn believed to have been killed and consumed by communal hunting tribes. Radiocarbon dating and the presence of other artifacts at the site suggest that the killing took place around eighty-five hundred years ago.

Farther south, in Nevada, there is a structure called the Fort Sage Drift Fence; it was constructed about three thousand years ago by a tribe skilled in organized hunting. The so-called fence is a series of rocks aligned for more than a mile that stand over three feet high in places. It is believed that the human builders used the concealment provided by the formation to ambush game, and that although pronghorn antelope (*Antilocapra americana*) were likely the main target, desert bighorn may also have been hunted.

the first to formally describe measurements and other features of the bighorn in scientific terms.

In 1805, shortly after Shaw's article was published, the Lewis and Clark expedition encountered bighorn for the first time. This sighting took place near the mouth of the Yellowstone River close to what is now the border of Montana and North Dakota. Joseph Field, the member of the expedition who first spotted the sheep, was fortunate enough to also find the horns of a bighorn that had died. He took the horns back to camp to show the other explorers.

Unregulated hunting

Exploration of the American West was followed by settlement. Faced with a rough environment and an absence of game laws, it was inevitable that the settlers would shoot bighorn for food. Scientist Warren E. Kelly commented: "The greatest hunting impact on the desert bighorn occurred during the nineteenth century, when the Southwest was settled by the eastern Americans. Common knowledge indicates the peak pressure was between 1850 and 1900."[3] Like the desert bighorn, the mountain bighorn more than likely faced peak pressure roughly during that same fifty-year stretch.

It is not known how many bighorn were hunted during the opening of the West. Today, conservationists and scientists monitor the bighorn sheep population, gathering data that can be used for analysis and comparison from year to year. This did not happen when the West was being settled. So while it is known that bighorn were shot and disappeared from a number of areas, particularly in the eastern portion of their range, it is uncertain how large a role hunting had in the wild sheep's demise. The impact of habitat loss, as the pioneers brought in domestic sheep and cattle and turned the former bighorn strongholds into ranchland, must have also been quite significant. What is known for certain is that the combination of hunting and habitat loss caused the disappearance of bighorn in many regions.

It is not surprising that the greatest hunting pressure occurred in the second half of the nineteenth century. This was the era when the frontier saw a constant influx of new

A ewe displays her radio collar, one method scientists use to monitor the bighorn population.

A trophy hunter takes aim at a bighorn ram with an impressive set of horns.

residents, game laws were minimal or nonexistent, and populations of bighorn were still sufficiently healthy to provide a substantial food resource.

As the years passed, bighorn became more scarce. Conservationists became concerned about the effect of hunting on wild sheep populations, so they began to exert pressure on western states to make laws against indiscriminate hunting. California acted early by banning bighorn hunting in 1873.

During the late nineteenth and early twentieth centuries, regulation of bighorn hunting in some states went through a two-stage process. The first stage was allowing bighorn to be shot, but only during a specified open season. The second stage was ending legal bighorn hunting entirely. For example, in Nevada, in 1895, a law was passed that declared bighorn could only be hunted from September 1 to

December 31. Populations of bighorn in Nevada continued to decline, however, and so in 1917 the shooting of bighorn was banned entirely.

Like Nevada, other states also saw bighorn numbers drop even with hunting regulations, leading to even more stringent regulations or outright bans on shooting wild sheep. Legal hunting of bighorn did not return to some western states until the early 1950s. By that time, many wildlife biologists were convinced that the bighorn population had recovered to the point that hunting could again be permitted. Some even argued that hunting might be necessary to prevent the overpopulation of bighorn in areas where they had thrived after a half century or more of complete protection.

Today, hunting is tightly regulated to ensure that shooting does not once again reduce the number of bighorn to perilously low levels. States have strict limits on how many hunting licenses are issued, open season is short (usually in November or December), and game lawbreakers are prosecuted.

The Impact of Early Hunters

It is not known for certain how great an impact hunting by pre-Columbian people had on bighorn populations. What *is* known is that the bighorn is one of the few large mammals in North America that escaped extinction at the end of the Pleistocene, the time period that encompassed the Ice Age. During the Pleistocene, North America was home to a number of species of camels, rhinoceroses, and wild horses. While these animals still exist in other parts of the world, they died out in North America between eight and twelve thousand years ago, as did the continent's saber-toothed cats, ground sloths, mammoths, and mastodons. Thus, many extinctions occured after humans had crossed into this continent from Asia.

Because of this timeline, some scientists have suggested that hunting by aboriginal tribes may have been partly responsible for the disappearance of these large mammals. If this is true, however, it raises the question why the list of animals that died out is so selective. Many tribes that hunted bighorn also pursued bison, deer, and other animals still alive today. If overhunting caused the Pleistocene extinctions, it is not clear why these animals survived while the camels, horses, and mammoths all perished.

Trophy hunting

To a nineteenth-century pioneer simply hunting to provide meat for his table, it mattered little what species of game he shot. Nor was the age or sex of the particular animal killed a major consideration.

Today the pioneer days are long gone, and so is the need to hunt for subsistence. Accordingly, over the past fifty years, hunters of bighorn have typically been trophy hunters, that is, they hunt to acquire an impressive-looking specimen whose head may be mounted on a wall. The desire for such trophies means that the most sought-after targets are rams eight years of age or older, because of their massive horns.

Simple supply and demand leads to a dilemma: There are not enough bighorn rams to be pursued by hopeful trophy hunters without a detrimental impact on the bighorn population. Thus, states initiated lottery systems in which a random drawing determined who would be issued one of

The beautifully spiraled horns of this ram make him a likely target for trophy hunters.

How Many Bighorn Were There?

It is only in the past few decades that dedicated wildlife biologists and volunteers have conducted painstaking censuses of bighorn. Likewise, it was not until the twentieth century that helicopters and other aircraft were available, invaluable not only for counting the bighorn but also for transporting researchers in roadless regions.

Because of these factors, it is not known with certainty how many bighorn sheep existed in North America in presettlement days. Thus, estimates vary enormously as to the size of the bighorn population before Americans moved west in the nineteenth century. In the 1920s, one scientist suggested that there were once 2 million bighorn in the United States. By the 1980s, estimates this high were viewed with suspicion, and it was written that there were probably never more than a half million wild sheep in all of North America; this estimate includes thinhorn sheep of Alaska and Canada as well as bighorn.

In the absence of reliable historical information, estimating the original population of any animal species is problematic. One possible way to do it is to figure how large the animal's original range was and calculate how many individuals such an area of land could potentially support. Unfortunately, this method assumes that the animal was uniformly distributed throughout that area of land, and it is therefore of limited use with an animal like the bighorn because of its selective habitat requirements.

the few licenses to hunt bighorn. A sportsman's odds against winning the lottery are as high as a hundred-to-one in some states.

Because of the long odds of these exercises in chance, a group organized in the late 1970s, called the Foundation for North American Wild Sheep, persuaded a number of game departments to donate permits to the organization to be auctioned to the highest bidder. The foundation believed auctions would provide a great deal of money that could then be used to fund bighorn conservation and restoration programs.

Auctioned hunts have indeed raised impressive sums. In 1995 and 1996, two permits to hunt bighorn in Alberta were sold for a total of $425,000 in U.S. dollars. A tire magnate from the Southwest paid $303,000 at auction in 1993 to shoot an Arizona desert bighorn ram. This same avid and wealthy hunter outbid others and spent a total of nearly $1.5 million between 1990 and 1997 for the privilege of hunting bighorn in several states. Buying a permit on the auction block comes with no guarantee; a hunter who paid $200,000 to shoot a thinhorn sheep in Alaska failed to bag his intended quarry.

The limited availability of bighorn hunting permits is reflected in the small number of animals legally shot. A good example is a population of bighorn inhabiting the Upper Yellowstone River valley in Montana. Over an eight-year period in the 1980s, this population consisted on average of twenty adult rams each year. During that same time frame, thirty-five rams were legally harvested, an average of only five rams killed per year. Even considering statistics for an entire state, the figure remains low. Nevada first permitted bighorn hunting in 1952, and over the first twenty-three years, 573 rams were shot. This number is fewer than twenty-five sheep killed yearly.

Conservationists naturally have a keen interest in tracking the number of bighorns harvested by hunters. Beyond simply gathering raw numbers, however, they seek to discover how hunting bighorn of certain age and sex classes impacts the entire population.

Effects of legal hunting

A theory lies behind the practice of trophy hunting. The classic notion is that by removing older rams, hunting gives younger males an opportunity to mate that they would otherwise lack; therefore, the population of bighorn is not negatively affected by hunting. In one sense the theory is inaccurate. By coursing, many younger male bighorn already successfully mate even if they are low on the hierarchy.

In a more broad view, however, the theory really is not about which males get to mate. It is about stability of populations. According to this theory, hunting does not decrease the number of bighorn because normal reproductive output compensates for the loss of the big rams. Thus, the number of younger males on hand to take the place of the harvested older rams should not be affected.

Again, this is not necessarily true. One investigation in Alberta indicated that when populations of bighorn lose a large number of rams five years old and older to hunters, there is often an increase in mortality of males age three to five years old. Scientists think this happens because when there are fewer older rams taking part in the rut, there are more fights between the younger males that lead to injury or death.

Scientists must deal with many intricate factors to determine how hunting affects bighorn populations. For instance, because it is a large mammal, the bighorn has a low reproductive output, so, to make sense, any study must be long term. Another consideration is that to be thorough, researchers need to consider whether the population could be affected by the hunting loss of individuals of any sex and age—not only big rams.

In Alberta, one group of zoologists attempted to alleviate these difficulties. They conducted hunting impact studies

The Ethics of Hunting

In the harvesting of bighorns by humans, there is a distinction between hunting, a legal sporting activity, and poaching, which is illegal. There are many people and organizations, however, who feel that even legal hunting is unethical or immoral.

Scientist and philosopher Ann Causey, in an article entitled "On the Morality of Hunting," noted that there are three general attitudes toward hunting prevalent in the general public. One group believes that all hunting is morally wrong on the grounds that it is cruel and inhumane and a violation of animal rights. A second group believes that hunting is wrong if it is done for recreation, but not if the hunter is simply seeking food for his table. Finally, there are those who think that all forms of legal hunting are acceptable and should be tolerated.

A study showed that city dwellers are far more likely than rural residents to object to hunting. Consequently, as the percentage of Americans living in urban areas rose during the twentieth century, so too did the percentage of hunting opponents. Recent surveys have shown that today at least half of all Americans had negative attitudes toward hunting. Only 7 percent of Americans today are hunters themselves.

Hunters protest that they should not be condemned, since their activities financially support conservation. Indeed, by some estimates the money paid for hunting licenses and other related fees provides over 75 percent of the budget for governmental wildlife management programs. As long as hunting persists, however, it seems likely that some opposition to the practice will continue.

that lasted more than twenty years. Furthermore, these scientists attempted to determine the effects of hunting ewes, not rams. They wanted to know whether harvesting ewes would cause a decline in the total population, whether it would affect the population of trophy rams, and whether it affected the size of a typical ram's horns.

To understand the efforts of these researchers, two aspects of their work must be considered. First, although these were hunting impact studies, what the process involved was in reality called experimental removal. That is, instead of waiting to judge the impacts of an actual hunting season, the scientists selectively removed animals from the herd to simulate a hunt. In this way, the scientists tightly controlled the number and the demographics of sheep that were "hunted."

Second, there is a reason why these zoologists guessed that there might be a relationship between the number of ewes and the horns' size in the first place. Most horn growth in male bighorns occurs during the animal's first seven years, and scientists believe that healthy horn growth is dependent on the quality and amount of food available to the young male. Thus, the zoologists thought, if the number of ewes was high, that would mean that the number of lambs would be high, too. More lambs means more competition for food and less food to go around in lean times. Thus, young male bighorn in crowded conditions might have diminished horn growth, which would mean that their horns would not be as impressively large when they reached maturity.

For nine years—a time the scientists refer to as the "removal years"—they kept the number of ewes stable by taking some of them out of the population. Then, the scientists stopped removing ewes. During these postremoval years, the number of ewes and yearlings tripled. If the theory was correct, male lambs born during the removal years would grow up to have larger horns than male lambs born during the postremoval years.

And indeed, this turned out to be the case; rams born during the removal years had larger horns. This is significant because, as noted earlier, rams with huge horns are the most sought-after trophies and the sale or auctioning of licenses to hunt them raises a great deal of money for wildlife programs. If a goal of bighorn management is to ensure that rams are of high quality as trophies, then according to the zoologists conducting the removal studies, it would be wise to harvest 12 percent of the ewes in the population annually. This would not decrease the total population nor increase the number of rams, but it would mean that adult males would have more impressive horns.

The scientists cautioned, however, that the results of their study in Alberta would not necessarily apply everywhere. If a population of wild sheep suffers heavy mortality because of predation or disease, hunting even 12 percent of the ewes could cause the total number of bighorn to decline. Thus, it must be stressed that where a decision whether or not to allow hunting is concerned, gathering reliable data on bighorn populations and demographics is essential. Only through careful analysis can wildlife biologists determine that a legal hunt, even one carefully regulated, will not negatively affect bighorn numbers.

Poaching

Although some authorities have suggested that legal hunting does not harm bighorn populations—and may even help them to remain stable—all biologists and conservationists condemn poaching. Poaching is the hunting of animals illegally. Legal hunting limits may be adjusted or hunting may be curtailed entirely based on data gathered and recommendations made by scientists. Poaching, however, takes place outside the law, so poachers do not reduce their kill based on restrictions imposed by state wildlife agencies. Therefore, among species of animals that have small populations, or that present a financial incentive for lawbreakers, the impact of poaching can be severe.

Poachers have been classified into three groups. The first group is accidental violators, those who through carelessness or error hunt an animal that is not legal game or is not in season. The second type of poacher is called a game

It is the job of game wardens such as these to prevent poaching of bighorn sheep and other protected species.

player; this person poaches simply for the thrill of trying to outwit game wardens.

The final category is the professional, one who kills animals illegally for the money that meat, hides, and horns will bring. This is probably the type of poacher that presents the greatest danger to the bighorn. On the black market, the mounted head of a bighorn may fetch more than $10,000, making this animal a particularly inviting target. As a trio of scientists concurred, "When wildlife products have a high market value, poaching becomes an increasingly tempting activity."[4]

Like legal hunters, those who poach bighorn typically do not seek ewes or young males. Instead, they pursue adult rams of at least four years of age, mindful of the ram's higher trophy value.

Impact of poaching

Zoologists conducting a poaching study of bighorn in Montana examined the records of law enforcement personnel over a recent six-year period. They learned that twenty-nine bighorn had been reported poached during that time. Nineteen of the illegal kills had been made by professionals; the others were accidents or the work of game players. Since twenty-nine incidents in six years is an average of fewer than five poachings a year, it might seem that the amount of illegal hunting was slight and the impact of the poachers minimal. However, poachers are seldom caught in the act and therefore many incidents go unreported. Accordingly, the researchers surmised that the actual number of bighorn poached must be significantly higher than merely twenty-nine animals.

No matter how many bighorn were illegally shot, raw numbers alone could not indicate whether these poachings had a significant impact on the population. Thus, the scientists formulated several theories of conditions that might be expected if indeed the level of poaching were high. For example, since trophy males are the poacher's quarry, it is likely that a high rate of poaching would mean that many mature rams were suddenly disappearing without apparent cause as they were quickly shot and hauled away. Also, if poaching were persistent, bighorn would be expected to become more wary and therefore should be less tolerant of humans approaching closely in vehicles or on foot.

After collecting data and using it to analyze their several theories, the scientists obtained inconclusive results. There was indeed a high rate of unexplained ram disappearances, which would suggest high poaching levels. On the other hand, bighorn in the study area often allowed humans to approach within about fifty yards, which would suggest poaching was not prevalent.

Since they could not exactly determine the extent of poaching's impact, the zoologists suggested that the wisest course of action was to continue the management policy already in place for bighorn. This policy includes educating people to obey hunting regulations and to report those who do not, and employing enough wardens to catch game lawbreakers.

Another cornerstone of bighorn management is to enact penalties for those convicted of poaching that are severe enough to discourage anyone from poaching in the first place. Penalties can include heavy fines or even imprisonment for repeat offenders.

Less research has been conducted on the impact of poaching on bighorn in Mexico than in Canada or the United States. From what is known, poaching seems to be a serious threat to Mexican bighorn, particularly since there are ap-

Although bighorn may reach speeds of thirty miles per hour, they usually try to elude predators by moving into rugged terrain where the predator cannot follow.

parently only about forty-thousand five hundred of the animals remaining in the nation. Mexico spends less on wildlife programs than its northern neighbors, thus there are fewer game wardens and other officials to deter poachers.

Conservation of bighorn is not simply a matter of deciding when and where they may be legally hunted. Humans impact the environment in so many ways that sometimes bighorn are affected indirectly by human activities. As with hunting and poaching, biologists must understand these factors to ensure a robust population of North American wild sheep.

3

Human Encroachment Threatens the Bighorn

BIGHORN, LIKE OTHER organisms, can only exist in environments that meet their biological needs. People alter the landscape by building roads, reservoirs, and cities, thereby eradicating natural habitats in favor of farms, pastures, and rangelands developed to support a growing human population.

Even in parks and refuges, where bighorn and other animals are fully protected, there may be a significant human impact from the large numbers of campers, hikers, and others seeking recreation who visit the parks to see the wildlife and natural beauty. Furthermore, sometimes even the most pristine sanctuary provides limited protection, because there might be dangers lurking for wildlife just outside the refuge's borders. Bighorn, like other animals, do not observe borders drawn on a map, and they may stray miles outside of protected regions.

Unlike hunting, where the impact on bighorn populations is intentional, most other means by which humans influence bighorn numbers are indirect. That is, bighorn are not purposefully harmed or put at greater peril. Even so, conservationists and scientists are concerned about these manmade potential dangers, which may include recreational activities, fire suppression, and diseases introduced into their environment by domesticated livestock. Because of this concern, a number of recent studies have sought to better understand the manner in which bighorn sheep may be affected by human encroachment in the American West.

Bighorn vs. nature lovers

Many of the outdoor hobbies that people engage in, such as backpacking or animal watching, are considered non-consumptive. This means that, unlike hunting, nothing is actually removed from the wild habitat. Still, even though backpackers and nature adventurers do not intend to have an impact on the environment, biologists wonder whether these human activities could have a potentially negative effect on wild sheep.

In the Sierra Nevada of California, scientists investigated the relationship between hikers and bighorn. They

Interaction with humans profoundly alters bighorn sheep behavior. Feeding bighorn or other wildlife is unwise and often illegal.

collected their data by interviewing hikers and observing interactions between bighorns and humans. They also conducted pellet transects, which means they counted bighorn droppings along lines of predetermined intervals. The study concluded that encounters between the bighorn and humans were limited to specific locations and that movements of the wild sheep were not adversely affected by hiker footpaths. Thus, bighorn residents and human visitors apparently coexisted amicably in these mountains, although the scientists cautioned that regulations limiting hikers to travel on designated paths should be maintained.

Dogs, authorities agree, should not accompany their owners on trails through bighorn habitats, because canine pets often chase or otherwise disturb wild sheep. Nor should off-road vehicles penetrate into critical habitat, since these too may distress bighorn.

Helicopter rides above the Grand Canyon offer breathtaking views but disrupt the feeding patterns of bighorn sheep.

Less encouraging than the Sierra Nevada project were results of a study conducted on bighorn and people in Grand Canyon National Park. Sightseers enjoy seeing the canyon from above, thus increasing the popularity of helicopter rides. Concerned that chopper overflights might disturb the bighorn, three scientists calculated time budgets—that is, how long bighorn spent on each of their daily activities—in situations where helicopters did and did not pass overhead. The investigation showed that while spring overflights had no significant effect, helicopters swooping overhead during the winter caused a 43 percent reduction in food intake. Since winter food may be scarce, such a drastic reduction in the amount of time bighorn can spend feeding may cause malnutrition or illness. The seasonal difference may have occurred because spring migration took the bighorn farther from the helicopters' flightpaths so that they were less likely to be disturbed by the aircraft.

The scientists calculated that helicopters needed to remain 820 to 1,500 feet away from the sheep to avoid disturbing them. The U.S. Navy has also adopted a 1,500-foot-minimum ceiling for military flights over bighorn habitat in southern California.

Fire suppression

Some former bighorn habitat is no longer suitable for these animals because humans have suppressed fires. Concern for protection of the forests led to extensive campaigns to eliminate all forest fires in many western parks. These efforts by foresters and rangers were successful, and the incidence of wildfire decreased significantly.

In the past few decades, as the ecology of western habitats came to be better understood, it was learned that preventing all fires was detrimental in many ways. Natural fires, started by lightning, are an integral part of nature's cycle of life in many western forests. They burn away leaf litter and debris and release nutrients into the soil, encouraging new growth by young plants. If every small fire is put out, leaf litter may accumulate to the point that seeds can no longer sprout. The excess debris can also transform the forest floor into a powder keg, in which a lightning strike that would normally start a small fire instead begins a massive one that engulfs hundreds of acres.

Another problem with fire suppression—and one that particularly affects the bighorn—is that without fire, forests become more dense as they expand to cover previously open landscapes with a plethora of trees. This creates conditions unfavorable to the bighorn, which tend to avoid dense forests for two reasons. First, bighorn prefer to stay in visual contact with each other as they traverse through their home range; this is difficult to do where vegetation is high and deep. Second, dense vegetation conceals predators such as mountain lions.

Because humans were so diligent about fighting fires in the past, much of the bighorn's former habitat has been rendered useless to them. An example is Mesa Verde National Park in Colorado, where fire suppression has caused at least 220 square miles of potential bighorn habitat to be no longer suitable for these animals.

Bighorn and man's animals

Humans also altered the West when they introduced livestock. Cattle and sheep now range in many areas that once were home to bison and pronghorn. These animals, as well as bighorn, have often been reduced or exterminated when humans appropriated their habitat for use by domesticated stock. When cattle and sheep are brought into bighorn habitat, they compete with bighorn for food.

The spread of sheep across the West has had another unintended effect on the bighorn. They may catch fatal diseases from their domestic cousins. Scientist Francis Singer and his colleagues summed up this problem by declaring: "The single greatest obstacle to the restoration of large, healthy populations of bighorn sheep in the western United States is epizootic outbreaks . . . that may kill 20–100%; of the animals in populations."[5] An epizootic is similar to an epidemic, except that while an epidemic is a disease that strikes a large number of humans, an epizootic is a sickness that strikes animals other than humans. Because of the dis-

Barbary Sheep

Because of the possible transmission of *Pasteurella* bacteria from domestic sheep to bighorn, wildlife biologists seek to minimize contact between the two species. Besides domestic sheep, bighorn have another relative in parts of the Southwest that could spread infection: the barbary sheep, or aoudad (*Ammotragus lervia*).

Barbary sheep are not domesticated; they are a wild species native to North Africa. In the late 1950s, they were released in western Texas and other arid regions that are now or were formerly inhabited by desert bighorn. It was hoped that the introduction of the barbary sheep would provide a new game animal for hunters.

By 1981, it was estimated that there were at least sixty-five hundred barbary sheep in the southwestern United States. Even if these exotic newcomers do not always pose a potential disease threat, they still often displace desert bighorn by competing with them for scant food resources available in arid habitats.

ease problems facing bighorn, veterinarians as well as biologists are involved in conservation efforts.

Infectious bacteria and viruses—or pathogens, as veterinarians may call them—are a part of nature, and usually healthy populations of animals can cope with their existence. In a stable population, diseases, as well as predation and other forms of mortality, do not remove from the group more individuals than are added by births. In an epizootic, however, this natural balance may be upset, causing an entire population of bighorn to vanish.

Chief among the illnesses bighorn may contract from domestic sheep is pneumonia, often caused by a bacteria called *Pasteurella*. In 1989, an article was published about the danger of bighorn coming into close contact with domestic sheep. A group of six young bighorns, five born in captivity and one taken from the wild as a lamb, were raised together in a five-acre pasture for more than a year. All six were examined for *Pasteurella* bacteria; none were found to have the bacteria.

At that point, six domestic sheep were placed in the pen with the bighorn. These domestic sheep appeared quite healthy, although four of the six did have some *Pasteurella* in their respiratory systems.

Within seventy-one days of the introduction, all six bighorn had died of pneumonia. Only one bighorn survived until the seventy-first day; the other five died within thirty-six days of exposure to the domestic sheep. On the other hand, the domestic sheep showed no signs of illness. Domestic sheep, it was discovered, can carry *Pasteurella* without contracting pneumonia, while bighorn's exposure to *Pasteurella* can quickly be fatal.

William J. Foreyt, the author of the article, issued a strong warning: "On the basis of [the] results of this study and of other reports, domestic sheep and bighorn sheep should not be managed in proximity of each other because of the potential fatal consequences in bighorn sheep."[6] Since the word "proximity" is vague, scientists and veterinarians recently suggested that to be safe there should be a gap of at least ten miles between bighorn and their domestic cousins.

Keeping bighorn and domestic sheep ten miles apart may seem a sound policy, but the bighorn themselves do not always cooperate. While female bighorn usually remain in an established home range their entire lives, bighorn rams may roam as they seek new groups of ewes to mate with. Their wanderings may bring them into contact with domestic sheep; they may even try to mate with their domesticated relatives. Nose-to-nose contact can infect the bighorn ram with *Pasteurella,* which he might then pass on to other bighorns—and thus an epizootic begins.

Lungworms, pneumonia, and limited space

Not only are pathogens an inevitable part of nature but so are parasites. Most animals are prone to harmful infestations of certain types of small organisms. These parasites are not to be confused with the beneficial microorganisms in bighorn stomachs that help the sheep to digest their food.

Although parasites are a natural part of life for most animals, they can be particularly troublesome when habitat reduction caused by human encroachment boxes animals into a smaller area than they inhabited before human contact. With suitable habitat reduced, animals may be restricted to small parcels of land, thus increasing the chance that parasites will be passed between individuals. This is the situation that may cause the bighorn to face a particularly great hazard from its most troublesome internal parasite, the lungworm.

The lungworm is a type of roundworm. Like many internal parasites, it has a regular, inflexible life cycle that requires it to infest two different species of animal hosts. In the lungworm's case, it needs a land snail as well as a bighorn to complete its life cycle.

The best place to begin a description of the lungworm's life cycle is where the adult worms live, namely, the lungs

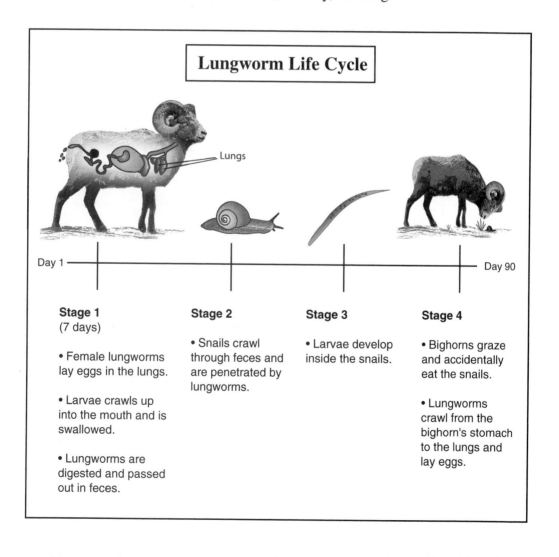

Lungworm Life Cycle

Lungs

Day 1 ——————————————————————————— Day 90

Stage 1
(7 days)

• Female lungworms lay eggs in the lungs.

• Larvae crawls up into the mouth and is swallowed.

• Lungworms are digested and passed out in feces.

Stage 2

• Snails crawl through feces and are penetrated by lungworms.

Stage 3

• Larvae develop inside the snails.

Stage 4

• Bighorns graze and accidentally eat the snails.

• Lungworms crawl from the bighorn's stomach to the lungs and lay eggs.

A wildlife manager administers lungworm medicine to a Rocky Mountain bighorn ewe.

of adult bighorn. Female lungworms lay eggs, which hatch within a week. These larvae then crawl out of the lungs, up the windpipe, and into the bighorn's mouth. Once there, they are swallowed by the bighorn and pass through the entire digestive system unharmed, finally passing out of the sheep in its feces.

Lungworm larvae continue the next stage of their life by penetrating the bodies of small land snails that are common in the low vegetation of bighorn habitat. Once inside the snail, the larvae develop in forty-five to sixty days into what are termed third-stage, or infective-form, larvae.

While grazing, bighorn often accidentally eat snails. When bighorn eat snails that contain infective-form larvae, these larvae travel from the stomach to the lungs. Once in the lung, they begin to develop into the adult stage, which takes about twenty days. Adult worms mate and deposit eggs, which then hatch to begin the cycle anew.

While adult bighorn often can tolerate lungworm infestations, the same is not true of the lambs. As many as 75 percent to 95 percent of the lambs in a population of bighorn may perish from pneumonia brought on by lungworms, and it is this high rate of mortality that ultimately affects bighorn numbers.

An adaptation of one of the two species of lungworms enables it to pass from a pregnant ewe into the embryonic lamb she carries. Scientists call this adaptation "transplacental transmission." Typically, this transmission happens in the last six weeks of the ewe's pregnancy. Once in the unborn lamb, the infective larvae enter its liver; when the lamb is born they migrate to its lungs. There, as in adult bighorn, they develop into adult, egg-laying lungworms.

It is in the lambs that a potentially fatal problem develops: If there are more than a hundred infective lungworm larvae, the animal often becomes susceptible to pneumonia. In some populations of severely affected bighorn, lambs may have more than 30 million lungworm larvae—and thus no chance of survival.

When massive numbers of larvae invade a lamb's lungs, they overwhelm the young bighorn's immune system as it attempts to fight the effects of the larval invasion. Thus, the infected lamb's lungs are susceptible to an invasion by pneumonia-inducing bacteria that could be warded off by a healthy lamb. The condition worsens: At about six weeks of age, lambs typically perish from an advanced condition known as verminous pneumonia.

The lungworm's life cycle results in a boom-or-bust cycle of bighorn populations. The more bighorn, the more feces they expel, the more larvae to infect snails, the more likely snails ingested by the bighorn are to have larvae, and the more larvae to pass from ewe to fetus. Thus, as the population of bighorn swells, the more chance there is that this population will actually crash and perhaps disappear entirely due to disease.

The classic example is the herd of bighorn at Pikes Peak in Colorado. In 1970, the population had grown to about a thousand animals. Many of these mountain sheep used the same areas for bedding most winter nights, thus feces were concentrated in these areas. When the bighorn left their winter range and moved uphill to summer ranges and lambing areas, all the feces in their winter "bedroom" caused vegetation to grow thickly and attract snails, which became infested with lungworm larvae. Fall came, and the bighorn migrated back to their winter range. The thick vegetation in the winter bedding areas was an irresistible food source, so the bighorn grazed, gulping down many infested snails in the process.

The result was that between 1970 and 1975, 95 percent of all lambs died soon after birth. With few youngsters be-

 Under Control

When veterinarians test the effect of a medication, they must be cautious before making conclusions about whether or not the medicine did what it was supposed to do. For example, suppose veterinarians give a group of bighorn a drug to prevent the herd from contracting lungworms. Inspecting the bighorn months later, they find few or no animals with lungworm infestations.

It would be easy to jump to the conclusion that the medicine worked. The problem is, How do the veterinarians know that these bighorn would not have contracted lungworms even if they had not received the drug treatment?

For that reason, veterinarians and scientists often use what is called a control group in their studies. For example, they would test at least two separate groups of bighorn, giving one herd the drug while the other herd—the control group—does not get the medicine. If the herd receiving the treatment does not develop lungworm infestations but the control group does, it is likely that the medicine worked. But if neither herd contracts lungworms, it is not certain whether the drug in this case had any effect.

ing added to the population and older animals dying off, by 1975 there were only 162 bighorn left whereas only five years earlier there had been a thousand. Since this epizootic was closely connected to aspects of the bighorn's own natural history, a team of scientists and veterinarians declared: "Their gregarious nature and marked tendency to develop patterns of habitat use make wild sheep their own worst enemy."[7]

Lungworm-pneumonia epizootics continue to pose a challenge today for wildlife biologists managing bighorn sheep. It should be emphasized that disease, like other natural factors, caused bighorn mortality before their environment experienced a human presence. What has changed is that because of human encroachment, bighorn inhabit vastly reduced amounts of land, increasing the likelihood that when epizootics strike, they will be more severe

Biologists employ a net to study a population of Rocky Mountain bighorn sheep dying from lungworm disease.

because the population is concentrated into smaller areas than in presettlement days.

Pneumonia and stress

In the mountain bighorn, lambs become susceptible to fatal bouts of pneumonia largely because of lungworms. But parasites are not the only cause of pneumonia infecting bighorn populations. In arid deserts where lungworms are not prevalent, desert bighorn, like their alpine cousins, also suffer from pneumonia, or, as it is also called, pasteurellosis. Pneumonia damages the respiratory system by attacking the lungs, trachea, and nasal cavities.

There are many different types of bacteria associated with pneumonia. One of them is called *Pasteurella,* from which the disease name pasteurellosis is derived. Bighorn are particularly vulnerable to *Pasteurella* and other pneumonia-inducing bacteria when they are fighting the effects of some type of stress, because stress weakens their immune systems.

These sources of stress may be natural. For example, heavy snowfall that makes it difficult to find food is a stress factor, as are parasite infestations. A number of human-

 Mineral Mortality

Recently, scientists have suggested an additional factor that may cause lambs to become susceptible to pneumonia—a lack of selenium in their diet. Selenium is a trace mineral, and only a tiny bit is needed by lambs to ensure normal growth and proper immune system development. Tests on bighorn food plants in Wyoming indicate that the level of selenium was 5 parts per billion; bighorn should get about 20 parts per billion selenium from their forage. A dearth of selenium in the plants eaten by ewes means too little of this mineral in the milk passed on to the lambs.

Scientists theorize that low selenium levels may be related to acid rain. Human agricultural and industrial processes have altered the composition of rainfall. The result is that the storms falling on the Rocky Mountains contain significant amounts of nitrates, ammonium, and other substances that may affect the soil's chemistry, in turn causing the selenium breakdown. Poor levels of selenium in the soil mean less selenium in the plants growing there. Thus, the bighorn diet lacks a sufficient amount of an essential component that helps to prevent disease.

caused circumstances can also lead to bighorn stress, such as harassment by dogs or large levels of noise associated with roads or construction. Another stress is overcrowding, where encroachment by civilization causes high bighorn population densities. Under stress, the level of steroids circulating in a bighorn's body increases, which decreases the animal's resistance to infection.

These conditions can produce pasteurellosis epizootics that have severe consequences. In the North San Juan District of southeastern Utah, there was a population of about 259 bighorn in the late 1970s. A pasteurellosis outbreak occurred there that wiped out the entire herd by the late 1980s.

Bighorn and botflies

There is not always a direct link between human encroachment and serious bighorn illnesses. Nevertheless, because preserving bighorn is a priority, conservationists have sought to understand diseases prevalent in bighorn so that these maladies will not further reduce bighorn numbers, especially in areas where bighorn numbers are already low.

One ailment, called chronic sinusitis, is more common in desert than in mountain bighorn. Again, a parasite is responsible. In this case, it is a common insect called the nasal botfly that depends on animal hosts to complete its life cycle.

The cycle begins when a female botfly travels to the muzzle of a domestic sheep or a desert bighorn. She deposits her larvae, which then crawl up the animal's nostrils and into its sinuses. It takes the larvae as long as ten months to complete their molts, after which they crawl back out the sheep's nostrils and are dispersed when the animal sneezes. After getting sneezed or snorted out of the host's body, the larvae turn into pupae and then adult botflies, beginning the cycle anew.

A heavy infestation of botfly larvae in its sinuses causes a bighorn to produce a large amount of mucous. This can lead to chronic fits of sneezing and head shaking as the animal struggles to expel the mucous. The bighorn may also stop eating.

Bighorn Parasites: Ecto and Endo

Parasitology is the study of organisms that live on or in another organism. The Greek word *ektos* means "outside," while the word *endon* means "within." From these roots come the words ectoparasite and endoparasite. The scabies mite that infests the bighorn's ears is an example of an ectoparasite; it attacks an outer part of the body. Lungworms, which conduct most of their life cycle inside the bodies of bighorns and snails, are an example of an endoparasite.

For the other prominent bighorn parasite, the botfly, it is less obvious which term should apply. After all, the adult fly does not live within or on the bighorn; it is free to move about on its own. On the other hand, its larvae live within the bighorn's sinuses. Because of the free-ranging nature of adult flies and because an infestation begins with a fly landing near a bighorn's nostrils, the botfly is considered an ectoparasite.

Sinusitis can eventually be fatal, with death resulting in one of several ways. As a result of the mucous buildup, infections may develop in the animal's head that lead to blindness, which makes the bighorn easy target for a predator. Eating may become difficult or impossible, causing some sinusitis-infected individuals to lose one-half of their body weight before perishing. Finally, large amounts of mucous in the nasal cavity may simply cause death from suffocation.

Veterinarians believe that desert bighorn are more susceptible to chronic sinusitis than their mountain relatives because the arid lands of the Southwest are better botfly habitat than humid mountain regions. Furthermore, since desert bighorn often come to specific places where water is available, botflies can congregate in these few spots and be assured of finding bighorn to host their larvae.

Mucous infections affect the bones in a bighorn's head, so veterinarians can inspect skulls of deceased bighorn to determine whether they suffered from chronic sinusitis. After examining 630 skulls, veterinarians concluded that 20 percent of them had this illness at the time of their death, and in many of these cases, sinusitis was judged to be the primary reason the animal lost its life. It was particularly common among bighorn ewes from Arizona—

45 percent of them suffered from the condition. Sinusitis mortality may therefore be a significant problem in places where bighorn numbers are already low.

Mites and scabies

Another widespread parasite that can be a significant threat to the bighorn is the scabies mite, an Arachnid. The Arachnids are a group of invertebrates that includes the predatory spiders and the scorpions. Unlike its relatives, mites are parasites.

Scabies mites primarily infest the ears of bighorn, causing the growth of numerous lesions, that is, wounds, scrapes, and scars. These lesions may become quite severe. One group of veterinarians theorized that the lesions make it difficult for bighorn to hear. To test their theory, they attempted to simulate the effect of a mite infestation in three captive bighorn by plugging their ears with bone wax or saline solution. They concluded that bighorn with scabies mites suffered a decrease in hearing sensitivity as a result of their infestation. This hearing loss could make a bighorn more vulnerable to predators, while the many lesions could render it more prone to infection.

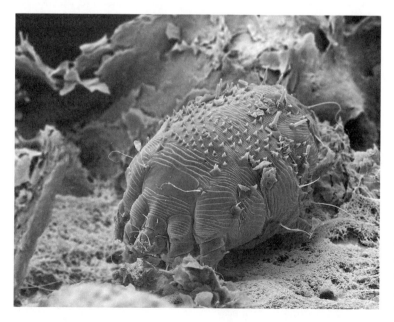

A magnified view of a scabies mite, an Arachnid parasite that poses a serious threat to the bighorn.

The death rate from scabies is not as high for bighorn as it is from pneumonia-associated epizootics. Nevertheless, severe outbreaks occasionally occur. In the San Andres National Wildlife Refuge of New Mexico, a scabies epizootic that began in 1979 is reported to have killed more than 90 percent of the park's bighorn.

Curing disease

Wildlife biologists seek to halt the decline of bighorn in areas where their numbers have become dangerously diminished. Since epizootics pose a particular threat to bighorn recovery efforts, the question has arisen whether bighorn can be medicated or inoculated against their most serious disease threats. Veterinarians have begun to address the problem, with varying degrees of success depending on the particular illness that is being fought. Sinusitis, for example, has thus far resisted effective treatment.

Somewhat more encouraging are the results of attempts to control pneumonia. When dealing with lungworm-associated pneumonia, the wisest course of action is to break the chain of the lungworm's life cycle at its weakest link. This, veterinarians have learned, is the point in the lungworm's life when it is stored in the lungs of ewes as infective larvae. At this time, the lungworm can be treated by oral drugs that destroy the larvae without hurting the bighorn.

The advantage of administering medicine orally is that the veterinarian need not touch the bighorn to treat it. In an area where pneumonia is a problem, typically apples are set out for the bighorn to eat. They may be suspicious of this new food source at first, but eventually they learn to eat the fruit without concern. Once the sheep are reliably eating their apple treats, medicine is mixed in. Thus, by chewing and swallowing apples, the bighorn medicate themselves. Furthermore, the drugs veterinarians use are quite safe, so there is little danger of a sheep getting a harmful overdose.

Recently, however, a team of veterinarians questioned whether these treatments are truly effective for diminishing the incidence of lungworm-associated pneumonia.

After conducting experiments that lasted four years, they concluded that even bighorn groups medicated for lungworms every year may still suffer heavy lamb mortality. On the other hand, some groups of bighorn ewes receiving no treatment at all for lungworms still gave birth to a reasonable number of healthy, surviving lambs.

Veterinarians had hoped that, as a preventative measure, bighorn could simply be vaccinated against *Pasteurella,* the bacteria most often associated with pneumonia. When animals—including humans—are given inoculations, the goal is for the vaccine to cause the animal to develop antibodies that will successfully fight the disease should the animal ever be exposed to it. A danger is that since the vaccine often contains a weak strain of the infective agent, it could ultimately cause the animal to catch the illness the vaccine is supposed to prevent. Thus, research is required to determine whether a particular vaccination is safe and effective.

Because of conservation concerns, testing *Pasteurella* vaccinations for bighorn is a priority project for many veterinarians. In 1999, a team of veterinarians reported that a new vaccine appeared to be safe for bighorn while providing at least some protection from the disease. Perhaps as the intricacies of the disease become better known, a vaccination providing nearly total protection can be developed.

It would be expensive and impractical to go into the field and capture all bighorn to vaccinate them. For the most part, bighorns vaccinated are those that are captured for other reasons, such as moving animals from one place to another in an effort to start new herds or augment existing ones.

Whether it is a case of sightseeing helicopters flying too closely to bighorn, suppression of natural fires, or domestic sheep bringing diseases into the environment, some human activities can unintentionally pose a threat to bighorn survival. Scientists and conservationists continue to monitor these threats and thus hope that they can productively manage the remaining bighorn population, in spite of the many problems caused by human encroachment.

4

Isolated Bighorn

IN MANY PLACES, bighorn sheep are confined to small pockets of suitable habitat. These pockets are often areas such as national parks and wildlife refuges in which the bighorn are fully protected. Nevertheless, these parks frequently are surrounded by ranches, highways, towns, or other landscapes maintained to support human populations. Bighorn cannot safely cross these wide areas of unsuitable habitat.

As a result, the population of bighorn is fragmented; that is, rather than having the nearly continuous range they may have had in the early nineteenth century, their distribution is piecemeal. Those wild sheep that live in a particular park are isolated there. If all the bighorn were to die out in a particular place, there is little chance that bighorn from elsewhere would migrate in to fill the void. Conservationists are concerned that the isolated nature of the bighorn's range, in conjunction with natural threats, might compromise bighorn preservation efforts.

Scientists also worry that since bighorn are confined to isolated pockets of habitat, they may have too few options when seeking to mate with other individuals. If all bighorn in a group are closely related, they will have no choice but to mate with their close relatives, leading to a condition called low genetic diversity.

Genes and genetic diversity

When the number of living individuals of a particular species of animal is small, there is a greater likelihood the animal will become extinct merely by chance. This is because the fewer the number of individuals that exist, the greater the possibility that some random event, such as disease or severe weather, could wipe out the entire population.

Furthermore, the fewer individual animals there are, the greater the chance that when an animal mates, it will do so with a close relative. This is called inbreeding. Inbreeding can be detrimental, because rather than receiving a mixture of diverse genes from two unrelated parents, offspring receive similar genes from the mother and the father. The result is a decrease in genetic diversity.

Low genetic diversity in a population is undesirable. High diversity of genes, on the other hand, is beneficial to animals. If offspring receive diverse genes, there is less

A solitary bighorn ram sniffs the surrounding air for a female. In isolated areas of habitat, bighorn may mate with their relatives.

chance that potentially harmful genes will be passed on to offspring. Harmful genes are also called deleterious genes. If these genes become common in a population, they can cause future generations to have undesirable characteristics that diminish their chance of survival. Also, if a population of animals have diverse genes, there is a greater chance that at least a few individuals will have characteristics giving them an ability to adapt to environmental changes or to resist illnesses. Natural calamities might claim the lives of other individuals who do not share the genetic makeup of the survivors.

Under conditions of low genetic diversity, a population is more likely to have deleterious genes present. It is also less likely that there will be random variations of genes, which would result in individuals more likely to be resistant to disease or other calamities. Low genetic diversity even affects animal fertility. Inbreeding can cause fewer young to be born, and those that are born are often not healthy.

Some scientists feel that the bighorn population is genetically vulnerable because it consists of small, scattered herds. These scientists fear that the bighorn's fragmented range makes the animal prone to local extinctions, that is, a tendency for all individuals to die out in one particular place. The more local extinctions that occur, the more the total population of bighorn in North America is reduced and made even more vulnerable.

The computerized bighorn

A living population of bighorn, however small and therefore vulnerable to extinction due to low genetic diversity, may take many years to actually vanish. Therefore, it is too time-consuming and expensive to monitor each stage in the group's demise. Furthermore, scientists and conservationists do not want groups of bighorn to actually go extinct; these people are in the business of trying to save the bighorn.

Computer models have been developed to try to predict how vulnerable bighorn are to local extinctions. In 1990, scientist Joel Berger created a model to simulate conditions that existed for 122 separate populations of bighorn

Captivity Closes the 125-Mile Gap

The bighorn population today consists of numerous small herds isolated from others of their kind. There is another type of isolation among North America's wild sheep, but this one is not influenced by human activities. It is the geographic isolation between the bighorn sheep and its cousin from Alaska and northwestern Canada, the thinhorn sheep.

The northern limit of the bighorn's range occurs in British Columbia. That province is also the location of the southernmost portion of the thinhorn's range. There is, however, a gap of about 125 miles between these two points, so bighorn and thinhorn do not associate in the wild. This natural isolation prevents North America's two wild sheep from interbreeding.

In captivity, animal species may be brought together that would not otherwise occur in the same place. Such was the case at the Yukon Game Farm in Canada where in 1994, a thinhorn ram mated with a bighorn ewe. They produced a female lamb that became a mother herself almost two years later. Her mate was another bighorn.

Since thinhorn and bighorn are able to interbreed in captivity, this suggests that they might do the same if they had overlapping ranges. If that happened, over time bighorn and thinhorn might "merge" into one species. As it is, they are kept separate by 125 miles of natural isolation.

residing in the Southwest. The model's analysis was striking; it predicted that 100 percent of the populations with fewer than fifty individual sheep would become extinct within fifty years.

These regional extinctions were predicted to occur even if there were no food shortages, harsh weather, predation, or competition from other species of animals. In other words, small populations of bighorn, according to the computer, would decline to extinction even in the best of circumstances, simply because when a population is so small, each successive generation will be less likely to produce healthy offspring because of inbreeding. Grimly summing up his model's predictions, Berger declared: "These data suggest . . . that local extinction cannot be overcome because 50 individuals, even in the short term, are not a minimum viable population size for bighorn sheep."[8]

Recently, some scientists have questioned whether extinctions of small populations of bighorns would occur as quickly as the 1990 article predicted. Scientist John Wehausen criticized the computer model because some of its predictions had not come true. For example, several herds of bighorn with fewer than fifty individuals, which the model predicted would go extinct, had actually increased to over a hundred animals. In summary, Wehausen wrote: "The model . . . does not support the strong population size effect on extinction probability it first appeared to provide."[9] There seems to be no argument, however, with the basic premise that a healthy bighorn population is best served by having large, genetically diverse herds located in many different places.

Founders and bottlenecks

Zoologists studying animal populations sometimes speak of founder size, which is the number of different individuals that were the ancestors of a population. Suppose six bighorn, three males and three females, are released in a region not already inhabited by these animals. Suppose further that these bighorn thrive and reproduce, and that their children and grandchildren also reproduce. Because of all these reproductions—or "recruitments," as scientists sometimes call them—the bighorn population in the area grows to one hundred animals. This may seem a significantly sized herd, but it began with only a few individuals, the six founders.

A closely related concept is called effective population size. Scientists calculate the effective size of a population by taking note of how many males and females of breeding age exist in the group. If in a particular locale, the effective population size is less than ten—that is, if there are fewer than ten bighorn of breeding age—that group of bighorn is said to be experiencing a severe bottleneck. A bottleneck refers to a situation in which the effective population is so small that low genetic diversity and severe inbreeding might threaten the group's continued existence.

An example of a bighorn population that zoologists are concerned about because of low founder size and low effective population size is found in Badlands National Park,

South Dakota. Before the mid-1990s, all the individual bighorn there descended from fourteen animals. As in many of the eastern portions of its range, the bighorn had been exterminated in the Badlands by the early twentieth century. In the 1960s, it was decided to reintroduce bighorn to the Badlands, and so a group of bighorn from Colorado were brought to the South Dakota park in 1964.

Rather than being released directly into the wild, the bighorn were confined within a 370-acre pen while they became accustomed to their new environment. Over the next three years, the twenty animals that were initially placed in the large enclosure reproduced, but many adults and lambs were lost to pasteurellosis. In 1967, when the bighorn were finally released, the herd numbered only fourteen individuals—the founders of the Badlands herd.

These animals thrived, and twenty-one years later the Badlands bighorn population was up to 140 individuals. In

Conservationists are working to keep the population of bighorn sheep as numerous and healthy as this desert herd.

the early 1990s, once again a pasteurellosis epizootic struck, and so by 1996 there were only sixty individuals left.

Although the Badlands bighorn herd had fourteen founders, the effective population size of the group was even lower, only six. This is because eight of the animals released were sons or daughters of the others. Since scientists believe that an effective population size of less than ten constitutes a severe bottleneck, the Badlands herd appeared to be in great danger of extinction because of low genetic diversity and severe inbreeding.

In addition to this low genetic diversity, the total population of sixty animals in the Badlands in 1996 also troubled scientists. A herd of fifty animals may not be a minimum viable population for bighorn even in the best of circumstances—and the Badlands herd was dangerously close to this level.

As a result of this information, a decision was made to increase both the genetic diversity and the total population

A view of the rugged landscape of Badlands National Park, where bighorn reintroduction efforts have been successful.

of bighorn in the Badlands by bringing in animals from elsewhere to augment the herd. Through the addition of animals from 1996 to 1997, the number of bighorn in the park was brought up to 160 individuals. There is room for still more, since scientists estimate that Badlands National Park has sufficient habitat to support a population of four hundred bighorn.

Tiburon Island bighorn

Concern about low genetic diversity of bighorn herds is not restricted to the Badlands. In the Gulf of California lies Mexico's Tiburon Island, or as it is known in Spanish, Isla Tiburon. Twenty desert bighorn founders were released on the island in 1975. By 1999, the population had risen to 650 animals. Overcrowding was a concern. Furthermore, conservationists wished to start new bighorn herds elsewhere in Mexico where the species had been exterminated in the past. Thus, bighorn were regularly captured on Tiburon and transported to other areas.

Recently, geneticists have been concerned because the Tiburon bighorn herd is descended from only twenty founders. The fear is that while the Tiburon population is large, the genetic diversity is low. Worse, since this herd has been used for many years as the only source stock for new bighorn herds in Mexico, a large percentage of the wild sheep population all over Mexico descends from those same twenty individuals.

It would be logical to conclude that the Tiburon bighorn had low genetic variation based simply on the low founder size. Nevertheless, scientists decided to gather data to determine whether they could confirm this premise. Blood samples were collected from the Tiburon Island bighorn and also from other desert bighorn found in Arizona. From these samples, genes were examined and compared. The Tiburon Island population had significantly less genetic diversity than its Arizona cousins.

Based on this information, the scientists strongly recommended that additional bighorn be released on Tiburon Island. These animals should be unrelated to the twenty founders brought there in 1975. The researchers also urged that the policy of using only bighorn from Tiburon Island to start new herds elsewhere in Mexico be changed. They advocated that these mainland populations be supplemented with individuals unrelated to the Tiburon founders.

Rams and gene flow

Conservationists working to preserve bighorn sheep in the Tiburon Islands, the Badlands, and elsewhere are learning how genetic diversity is maintained in normal, healthy bighorn populations. Research has demonstrated that the rams are the primary agents in gene flow.

This is because female bighorn often spend their entire lives on the same home range where they were born. Thus, even in a purely natural state without human interference, the behavior of the ewes could cause low genetic diversity. Because rams seek out new breeding opportunities away from their birth site, their behavior acts as a deterrent to severe inbreeding. Because of this behavior, wildlife biologists are interested in the movements of these rams through the landscape.

A bighorn ram intent on seeking ewes might travel to another place that is miles away, oblivious to any obstacles in his way. Conservation challenges arise when there are manmade hazards in the roaming ram's path.

The problem is well demonstrated in southern California. Male bighorn have been reported wandering through the streets of downtown Palm Springs. However, female bighorn in southern California typically do not engage in forays into urban or developed areas. In fact, the home range boundaries of several ewe groups coincide with roadways that they seldom cross, if ever.

A striking contrast in the behavior of rams and ewes where roads are concerned is provided by California State Highway 74, south of Palm Springs. In the 1970s, ewes were sometimes seen crossing this road. Between 1970 and 2000, however, the amount of vehicular traffic on Route 74 tripled. As a result, ewes apparently began to avoid the highway entirely. Between 1993 and 2000, several ewes

The jumping prowess of this bighorn ram demonstrates how difficult it is to confine the movement of the male of the species.

from both sides of Route 74 were fitted with radio collars and their movements monitored. None crossed the highway. Probably in the 1970s and earlier, Route 74 simply happened to lie within the home range of one or more ewe groups, but as traffic increased, these herds adjusted their activities so that the highway is now the border of their range and not a line through it.

In contrast to the ewes, between 1991 and 2000, at least five rams were struck by vehicles as they attempted to cross Route 74. Two of the males were killed in these collisions.

Construction of highways such as California's Route 74 can severely impact bighorn movements, particularly if traffic is heavy. The genetic well-being of the bighorn population may suffer because rams can no longer travel safely from ewe group to ewe group. As for the ewes, they too may be affected. Although they do not range as widely as the rams, they still require a home range with good areas to forage for food and to give birth to offspring. Highway construction and traffic increases can cause ewes to shrink their range and lose choice bits of habitat.

A herd of bighorn sheep crosses Canada Highway 1. Roads and highways often cut straight through the territory of the bighorn.

The rams meandering through downtown Palm Springs were probably following a natural movement pattern that enabled them to mate with females in a number of different canyons and mountains throughout southern California's Peninsular Ranges. Roads, fences, canals, and other human constructions may hinder these rams and lead to a bighorn population with far less genetic variation.

The consequences of inbreeding

Computer models and established genetic principles give zoologists valid reason to believe that the low genetic diversity of many bighorn herds is detrimental. It is not as easy, however, to cite examples where there is physical evidence that the suffering of a bighorn population is related mostly or entirely to genetics.

In one sense this is to be expected. Suppose, for example, a severe epizootic strikes a herd of bighorn that has few founders and low genetic variation. Suppose further that the disease kills 80 percent of the bighorn in this population. Perhaps the epizootic would have been less severe if the group had more founders. Maybe if that had been the case only 60 percent of these bighorn would have perished. But there would be no way to know this for certain. Scientists cannot create a herd of bighorn in the laboratory to duplicate the conditions bighorn experience in the real world.

One place where there may be physical evidence of the effects of bighorn inbreeding is at the National Bison Range in Montana. The bighorn herd there was started with a small number of founders brought from elsewhere. In 1983 this herd numbered forty-seven animals, with fifty-one the following year. Several of the lambs born during those two years were underweight and very weak, leading zoologist Christine Hass, who conducted research at the Bison Range, to suggest that heavy inbreeding in this population was responsible for the poor conditions of the lambs. These lambs were heavily preyed on by coyotes, so it is possible that their weak condition made them particularly vulnerable to predators.

Predation is a natural factor. A number of carnivorous animals began to prey on bighorn, long before humans set foot in North America. In a natural setting, predation on bighorn and other large animals helps to prevent their numbers from increasing to a point where there are more bighorn than the habitat can support.

Today, because fragmentation of bighorn habitat has reduced the number of wild sheep in many areas to small herds, zoologists are concerned that predation may no longer benefit bighorn. They fear that a normal level of predation, or a particularly efficient predator, might undercut efforts to conserve bighorn. In recent years, scientists have conducted a number of studies to better understand the role predators play in the current status of North America's wild sheep.

Tracking bighorn

One way zoologists learn what carnivorous animals prey on is to be fortunate enough to see the predator in the act of making a kill. Because of eyewitness accounts, it is known that bobcats (*Lynx rufus*) prey on young bighorn.

Even when these reports are well documented by reliable observers, however, anecdotes are not an efficient method for learning about predator-prey relationships. For one thing, these observations are too dependent on luck. For another, based on a single observation of a bobcat killing a bighorn, a scientist cannot determine whether such behavior is common or rare, whether bobcats usually prey on some other kind of animal or not, or whether bighorns face more formidable predators or not.

What is needed is a more thorough and systematic way of studying bighorns and their predators. Fortunately, radiotelemetry can provide the comprehensive data zoologists need. In radiotelemetry, a collar containing electrical circuitry is placed on an animal. These circuits emit pulses that researchers can detect with a monitoring device at distances as far as twenty miles away. Placing radio collars on bighorns has shown wildlife biologists that, in some areas, the greatest natural enemy of desert and mountain bighorn alike is the mountain lion (*Puma concolor*).

With radiotelemetry, zoologists are alerted to likely events in the marked animal's life by its movements—or even by its lack of movements. If a researcher monitoring signals

An adult ram bolts down a hillside with his greatest natural predator, the mountain lion, following close behind.

One Predator, Many Names

The large cat of the American West that preys on bighorn sheep is called either a mountain lion or a cougar. Because this carnivore has such a broad range—from Canada down into South America—it is not surprising that it has a number of different common names. This cat is called a panther in Florida and a puma in some parts of Latin America.

However numerous its common names, like all living organisms it has only one official scientific name, which in the mountain lion's case is *Puma concolor* (although some older books and journal articles use the name *Felis concolor*). Part of the value of scientific names is that they are consistent across geographical regions and international boundaries, while common names often are not.

notices no change in a particular animal's location over a certain period of time, he or she might travel to the signal's source to learn what has happened. Perhaps the transmitter has simply fallen off the animal, or perhaps the animal has perished. Some radio collars even have automatic switches that cause the pulse rate of the signal to quicken if the transmitter is motionless for more than two hours.

In either case, by going immediately to the location of the signal, the scientist can salvage the lost radio collar— or discover what caused the death of the individual animal that was wearing it. The more radio-tagged animals that are monitored, the more data zoologists obtain on cause of mortality when death eventually comes.

Predators

Because of radiotelemetry, a recently published study on bighorn mortality contains enlightening observations. Scientists in southern California placed radio collars on 113 desert bighorn sheep. They monitored the sheep from the fall of 1992 through the spring of 1998. During the study period, sixty-one of the tagged bighorn died. Of these, forty-two were killed by mountain lions, which means that the big cats were responsible for 69 percent of all bighorn mortality.

Another study of mountain lion predation on bighorn was conducted in Alberta, Canada. In contrast to the California research, the mountain lions wore transmitters in this study, not the sheep. The investigation focused specifically on predation by the cats during the months of November through April—the long alpine winter season—over a period of nine years from 1985 to 1994. Sixty mountain lions were collared, and over the study period the zoologists examined 320 remains of animals the cats killed. The bulk of the mountain lion diet was mule deer (*Odocoileus hemionus*), with 183 kills. By contrast, only twenty-nine mountain sheep were preyed upon, of which eleven appeared to have been disabled or injured in some way. Thirteen of the killed bighorns were lambs, nine were ewes, and seven were rams.

Significantly, the study found that bighorn kills were not randomly distributed among the mountain lion population. Instead, only a few particular cats sought sheep as a key part of their diet. Over the course of one winter, a single female mountain lion killed eleven bighorn, including six lambs.

After mountain lions, the most significant predator of bighorns is probably the coyote (*Canis latrans*), although they are more of a threat to lambs than to adults. At the Na-

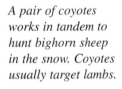

A pair of coyotes works in tandem to hunt bighorn sheep in the snow. Coyotes usually target lambs.

tional Bison Range in Montana, a herd of about fifty bighorn produced twenty-three lambs in the spring of 1984, but only two of these youngsters were still alive by mid-September. Coyote predation caused the most deaths, and more than two-thirds of the preyed-upon lambs had been taken by the coyotes within three days of their birth.

Wolves (*Canis lupus*) also prey on bighorn, although sparingly. In Banff National Park in Alberta, seventy-one wolf kills were examined by zoologist David Huggard. Most victims were American elk (*Cervus elaphus*) or deer; only eight bighorn were killed. Huggard thought that this prey choice occurred simply because wolves usually frequent habitats where they are more likely to encounter elk and deer than bighorn. However, wolves will kill bighorn if they encounter them while hunting.

The isolated nature of their populations makes each particular group of bighorn more susceptible to extinction than would be the case if these animals still inhabited the extensive range they held before human encroachment. Although human activities in the West have been responsible for isolating the bighorn, humans may also hold the key to preserving the United States' wild sheep through different conservation methods.

A gray wolf feeds on a bighorn ram. Although the wolf's diet generally excludes bighorn, the predator is not averse to killing sheep encountered while hunting.

5

Bighorn
Conservation

A GOAL OF managing threatened or endangered species is to raise their numbers to levels where they are no longer in peril. Scientists, conservationists, and volunteers are involved in many projects to ensure the bighorn's survival. Some of these projects involve habitat preservation and restoration, captive breeding, public education, controlling predators, and translocation.

Translocation

In the early twentieth century, when legal bighorn hunting ceased and protected refuges were established, conservationists believed that they could help bighorn survive by initiating translocation projects. This means that bighorn were captured in one area where their numbers were secure and shipped cross-country to places where they either were present in very low numbers or had gone extinct.

The extent of these translocation projects can be seen today by examining census figures for bighorn in the western United States. Data from 1990 indicated that Colorado's bighorn were distributed among sixty-eight separate populations, or herds. Of these sixty-eight herds, forty-three were transplants; that is, their origin can be traced to bighorn brought into the area by wildlife biologists. In several states, all the bighorn are the result of translocations.

Oregon, for example, had twenty-five herds in 1990; all of them were transplants.

When translocations began in the 1920s, bighorn biology was poorly understood. The notion that translocation would be a simple management tool with a high degree of effectiveness was soon proved wrong. While some translocations were successful, many failed.

As research was conducted and data accumulated, scientists began to put together a clearer picture as to what made translocations succeed or fail. One group of zoologists studied one hundred translocations that had taken place in six western states between 1923 and 1997. These scientists classified translocations in one of three categories. If the number of bighorn remaining at the site where animals were released was twenty-nine or fewer, the attempt was considered to be unsuccessful. A translocation was judged moderately successful if the new population had thirty to ninety-nine individuals. Finally, a successful project resulted in a population of one hundred or more bighorn.

A team of workers from the Wyoming Game and Fish Department captures a bighorn ram as part of a translocation project.

Counting at the Waterhole

At southern California's Anza-Borrego Desert State Park, bighorn have been censused annually since 1971 by a method called waterhole counts. In this arid environment, bighorns regularly visit the few reliable water sources, which allows them to be surveyed by scientists and volunteer counters.

There is a formal procedure for making a count. First, observers at a station must be as quiet as possible. All bighorn that come to the water between 7 A.M. and 5 P.M. for two consecutive days and from 7 A.M. to 2 P.M. on the third day are tallied.

Scientists want to learn as much as possible about the natural history of peninsular bighorn, so waterhole counts give them an opportunity to gather more data about the animals than simply their raw numbers. For this reason, observers are asked to record the age and gender of the individual sheep and also note whether any of them are marked animals wearing radio collars. In addition, observers record any noteworthy behavior, such as mating or nursing by the lambs.

To be most effective, waterhole counts must be conducted at the hottest and driest time of the year, since that is when the bighorn are most likely to all be drawn to the water. Traditionally, the bighorn counters conduct their census during the Fourth of July weekend.

With this standard in place, the team sorted the translocations. Of the one hundred projects, thirty were classified as unsuccessful, twenty-nine as moderately successful, and only forty-one as successful. Adding unsuccessful and moderately successful projects means that 59 percent of the translocations failed to bring the number of bighorn up to a level where long-term survival of the new population was likely.

After classifying translocations, the scientists conducting this study analyzed all the various factors that might explain the success or failure of each attempted project. First, they concluded that translocations were less likely to be successful when domestic sheep were located within 3.7 miles of areas the bighorn used.

Second, translocations had a lower chance of success if domestic cattle grazed in the same range as the bighorn. While cattle pose far less of a disease danger to bighorn than domestic sheep, they are a competitor for food.

Finally, it was determined that translocations were twice as likely to be successful if indigenous herds were used as the stock for an attempted reintroduction. An indigenous herd is one brought to the release site from nearby. If an attempt is being made to establish a new population in western Colorado, for example, it is far better that the bighorns used as the "seeds" for this population also come from Colorado instead of Alberta. Indigenous herds are already familiar with the vegetation and climate of the area, and they are more likely to settle into a seasonal migration pattern that fits well with their new home.

In some cases, use of indigenous stock for transplants is impossible. In states such as the Dakotas and Washington where bighorn had been completely exterminated, there was no choice but to bring in herds from considerable distances away for translocation projects.

Bighorn meet the GIS

Translocating bighorn is a big project that requires the efforts of many people. It is also very expensive. Because of the costs involved, research has been concentrated on predicting the success of bighorn translocations before a single animal has been captured or moved. Computer programs called geographic information systems (GIS) have made their way into the conservationist's toolbox as a means of forecasting the potential success or failure of a proposed translocation.

On a GIS program, an area of land appears as a digital map on the computer screen. Horizontal and vertical lines are projected across the screen so that the map is divided into a grid consisting of numerous cells of equal size.

In a typical GIS application, data is entered about a variety of bighorn needs and hazards, and the computer is instructed to highlight all cells that meet certain criteria. Because of this feature, a GIS program can provide conservationists with information to determine whether it makes sense to attempt to translocate bighorn to a particular place.

For example, a team of zoologists recently used GIS to answer the question: Should bighorn be translocated into

Mesa Verde National Park? Located in southwestern Colorado, Mesa Verde historically had a healthy population of bighorn, but by 1997 the number had dwindled to no more than four individuals. Since translocations are expensive, it would not make sense to expend resources by putting more bighorn in Mesa Verde unless there was a good chance that the project would succeed.

Using all the relevant data entered into their computer, the scientists first considered the bighorn's need for rugged terrain to avoid predators. Thus, the team identified all

In parts of the west, forests are too dense to provide suitable bighorn habitat; thus, many places are unsuitable for bighorn reintroductions.

The Safest Way to Catch a Bighorn

Some important aspects of bighorn management involve capture. Individual bighorns may need to be caught so that radio collars may be installed or repaired, signs of disease may be noted, or particular individuals may be loaded into crates to translocate elsewhere.

Veterinarians involved in these procedures naturally want to know what method of capture is least stressful on the bighorn. Between 1980 and 1986, 634 bighorn were captured in the western United States using one of four methods: chemical immobilization, dropnet, drive-net, and net-gun. The number of sheep caught by each method varied; for example, 249 were captured by drive-net while only ninety were immobilized.

Chemical immobilization means that the animal was shot with a dart containing a tranquilizer. Veterinarians concluded that this was the most stressful capture method and should not be used unless absolutely necessary. Net-gun—where a small explosive charge sends the net over the heads of the bighorn—was judged the least stressful and thus the best capture method. Nets dropped over the heads of bighorn from a supporting structure above (drop-net) or stationary nets that the bighorn were herded into (drive-net) were judged a better choice than chemical immobilization but not as safe for the sheep as net-guns.

areas on their GIS map with steep slopes, or located within a close distance to such slopes. By this standard, Mesa Verde was found to have 243 square miles of potentially suitable bighorn habitat.

Since bighorn cannot live on highways, the scientists deducted from the amount of suitable habitat the area occupied by roads. As a result, the 243 square miles of habitat dropped to 240 square miles. Discarding the areas that were heavily used by human visitors to the park brought the number down a little more, to 239 square miles. As the scientists added various factors, the amount of good bighorn habitat further decreased. This is typical in projects using a GIS system; subtraction takes place until all criteria have been taken into account.

The scientists next eliminated from consideration any area having what they considered poor horizontal visibility, that is, places where the vegetation was too tall and thick to be used by bighorn. When this factor was considered, the figure dropped all the way down to less than sixteen square miles of suitable bighorn habitat.

The team gave the computer three more factors to analyze, each reducing the estimated prime bighorn habitat even further. The result was that after taking these seven steps, the computer showed that Mesa Verde National Park had less than one square mile of ideal bighorn habitat. "This is insufficient area to support a viable population of bighorn sheep,"[10] the scientists determined in their report. Their conclusion was that bighorn should not be translocated into Mesa Verde but instead into five other western national parks where conditions were more favorable.

This is not to say that translocating bighorn to Mesa Verde is out of the question forever. In the future, controlled fires might thin out the dense vegetation that caused the GIS to reject potential bighorn habitat.

The peninsular bighorn

Perhaps in no other place have bighorn sheep garnered as much attention recently as in southern California. In a cluster of uplands and canyons collectively known as the Peninsular Ranges, a population of 334 desert bighorn lives within a two-hour drive of the millions of people living in the Los Angeles and San Diego metropolitan regions.

The state of California first designated peninsular bighorn a rare animal in 1971. Later, in 1984, the designation was changed to threatened. Finally, in 1998, the U.S. Fish and Wildlife Service listed this population as endangered.

In 2000, the U.S. Fish and Wildlife Service published the Peninsular Bighorn Recovery Plan. This document describes in detail the biology of southern California's bighorn, the threats they face to their survival, and how their future can be secured.

Some of the serious threats to bighorn in other places, such as crippling epizootics and poaching, are not prevalent in southern California. Instead, the peninsular bighorn has been placed at risk mostly because of manmade alterations to its habitat. The recovery plan said the following about habitat loss:

It represents a particularly serious threat to Peninsular bighorn sheep because they live in a narrow band of lower elevation habitat that represents some of the most desirable real estate in the California desert and is being developed at a rapid pace.[11]

Often it is not simply the presence of humans but rather the introduction of human artifacts that harms peninsular bighorn. For example, in the Santa Rosa Mountains south of Palm Springs between 1991 and 1996, five bighorn perished because of automobile collisions while another strangled itself in a wire fence. Five others were killed by plant poisoning after consuming leaves of laurel cherry or oleander, old-world shrubs not native to California but commonly planted as ornamentals. Even swimming pools may be fatal; there have been cases of lambs falling into pools and drowning.

Peninsular bighorn conservation

The recovery plan of the U.S. Fish and Wildlife Service proposes a number of actions to address threats to the peninsular bighorn's survival. The first recommendation is to promote population increase and to protect habitat. Protecting habitat means more than simply saving the remaining bighorn strongholds from development. It is also endorsing a policy of restoring and enhancing the lands already set aside for bighorn. An example of enhancing habitat for bighorn is removing stray cattle from the bighorn's habitat, since these domestic animals compete for scarce food and water resources in these arid lands. At Anza-Borrego Desert State Park, which is home to about 180 bighorn, park employees removed 117 cattle during a two-year period from 1987 to 1989.

There are several ways to promote population increase. One viable option is through captive breeding. In 1982, a nonprofit organization called the Bighorn Institute was established. The institute maintains a captive herd of about thirty peninsular bighorn in large pens at Palm Desert, California. By not allowing these young bighorn to grow

Oleander, a shrub commonly used in ornamental landscaping, can prove fatal to bighorn sheep if ingested.

accustomed to humans, the institute insures that when these animals are released into the wild, they will still have some fear of people and will hopefully avoid towns and other developed areas where they would be at risk. Between 1985 and 2000, seventy-seven captive-raised bighorn were released into the wild by the Bighorn Institute.

Another part of the far-reaching recovery plan proposal is to support research programs necessary for the recovery efforts. Examples of this type of process include taking a census of bighorn through helicopter surveys for waterhole counts, where the number of animals coming to water sources to drink is tallied. Knowing how many bighorn exist, how and where they are distributed, and the ratio of rams to ewes is basic information needed by biologists plotting conservation strategies.

A final component of the recovery plan is to promote education and public awareness programs. As the plan notes:

> Conservation efforts have a higher chance of success if they are supported by the local community. A number of recovery actions outlined in this recovery plan will directly affect the general public. . . . The public

needs to be informed of the reasons why specific recovery actions are being taken. [12]

Education of this sort can take many forms. Bighorn conservationists may visit schools, senior citizen centers, or social clubs in southern California to show slides or videos and to answer questions. Or, people could be taken on field trips into bighorn habitat. This might inspire them to participate in projects, such as waterhole counts, that depend on volunteers.

The road to successful recovery

The cost of performing all the actions recommended in the Peninsular Bighorn Recovery Plan is estimated to be about $73 million. The cost is spread out over time, however, since it is a long-range plan that will take at least until 2025 before its goal is met. That goal is to have the population be large and diverse so that the peninsular bighorn would no longer be considered an endangered species.

Many Human Friends

The Peninsular Bighorn Recovery Plan lists many people as members of the recovery team. These concerned individuals work to support peninsular bighorn restoration projects, and they represent a variety of organizations. Some work for federal agencies such as the U.S. Forest Service, the Bureau of Land Management, and the U.S. Fish and Wildlife Service. Others are affiliated with state organizations such as the California Department of Fish and Game and the California State Parks. Still others work for the University of California at Davis.

Representatives of private groups are also on the team. They are from groups like the Bighorn Institute and the Zoological Society of San Diego. Finally, a tribe of Native Americans, the Agua Caliente Band of Cahuilla Indians, also have a member on the board. The peninsular bighorn's habitat, or potential habitat, includes portions of three Indian reservations.

The diversity of the recovery team reflects the diversity of the bighorn's range. It extends across federal, state, tribal, and private lands. It is only by the cooperative efforts of a number of organizations and a number of dedicated people that the bighorn survive.

The removal of an animal from the endangered species list is called delisting. However, there is a step that must be taken before delisting called downlisting, which means the species is moved from the U.S. Fish and Wildlife Service category of endangered to the less perilous designation of threatened.

The recovery plan specifies standards that must be met before the peninsular bighorn may be either downlisted or delisted. To be downlisted—considered threatened rather than endangered—the major requirement is that nine separate regions of the Peninsular Ranges must each have at least twenty-five bighorn ewes. That means there must be a minimum total of 225 females for the entire population. Furthermore, the minimum of twenty-five females in each of the nine ewe groups must persist for at least six consecutive years, which is considered one bighorn generation.

A ram scratches his head against a hillside. Despite successful recovery efforts, bighorn populations remain vulnerable.

Also, adding captive-bred animals to herds would not count toward reaching these numbers.

Since there are currently more than three hundred peninsular bighorns, at first glance the delisting standard of 225 ewes might seem easy to achieve. However, the other requirement is that these animals must be distributed among nine particular areas. This criteria is nowhere near being met. One of the nine specified regions is the northern Santa Rosa Mountains. In 1998, there were only ten ewes in this area, and six of those had been raised in captivity. Clearly, the peninsular bighorn population needs to grow much larger before it can be downlisted.

To be delisted, or secure to the point where it could not be considered threatened, even more stringent standards must be achieved. The scientists would need to determine not only that there were twenty-five or more ewes in nine separate regions but also that the total population of peninsular bighorn averaged 750 or more individuals—rams, ewes, and lambs—for a period of at least twelve years. Given the current population and the bighorn's low rate of reproduction, the recovery plan will take at least twenty-five years to ensure the peninsular bighorn's continued survival.

While the specific conservation needs of southern California's bighorn have commanded a great deal of attention, a number of policies have been proposed to better secure the bighorn's future throughout its range. Some of the suggested management schemes do not please everyone.

Conservation and controversy

Certain proposals to help bighorn are met with little or no protest, while others are quite controversial. An example of a bighorn management project that does not provoke much dissent is the effort to control tamarisk, or salt cedar, an exotic plant introduced into the Southwest for use as hedges or ornamental plantings. This plant is very aggressive and can easily displace the vegetation that bighorn feed on. Perhaps worse, tamarisk consumes a great deal of water; if it starts to grow near the waterholes the bighorn visit, it can suck these areas of standing water dry.

In many places, efforts are being made to eradicate tamarisk, either by digging up plants or by careful application of herbicides. Since the plant is not native to the Southwest, and since water is already scarce in the region without the burden the tamarisk adds, most people agree that controlling this aggressive invader is sound policy.

However, other potential means of helping to stabilize or increase bighorn populations incite much debate and disagreement. A notable example is the suggestion that bighorn populations could be stabilized or enhanced through removal of mountain lions, particularly individual cats that have demonstrated a tendency to hunt bighorn. When bighorn numbers are normally high, this natural mortality is not harmful to the overall population.

Where bighorn are restricted in number or habitat, however, mountain lion predation can have a severe impact. For example, one herd of twenty-five bighorn in New Mexico was reduced to a single surviving member within eighteen months because of mountain lion predation. Because of incidents such as this, the question is often raised whether some of the big cats should be killed to help preserve the bighorn.

In 2001, the Arizona Game and Fish Department proposed killing about a dozen mountain lions roaming the Tonto National Forest east of Phoenix. The department specified that the plan's goal was to increase the number of bighorn in the area, which had dropped 65 percent between 1994 and 1998.

The proposal was both supported and condemned. Gary Barcom, the vice president of the Arizona Desert Bighorn Sheep Society and a supporter of the plan, said: "It will have a short-term negative impact on the lions, but a long-term positive impact on the sheep and the lions. As unpleasant as that type of management may seem, it's something necessary to bring the two into balance so they can coexist."[13]

Sandy Bahr, the Arizona conservation chairperson of the Sierra Club, disagreed. "The proposal doesn't have any sound basis in science," she declared. "They'd be killing off the mountain lions for no reason. Predators and prey

A mountain lion savors a fresh bighorn kill. A controversial program to stabilize the declining bighorn sheep population involves killing mountain lions.

have evolved over thousands of years. Just because you get rid of the predator doesn't mean you'll improve the population of the prey."[14]

Complicating the issue further, in parts of California the concern is that mountain lions pose a threat not only to bighorn but also to humans. In 1994 two women were killed by these big cats; another woman was similarly slain the following year. California had given mountain lions full protection in 1972. As a result, the number of these big cats in the state rose over the next twenty-five years from an estimated twenty-four hundred to six thousand. All

How Many Bighorn Are There Today?

It was estimated that there were between sixty-five thousand and seventy-one thousand bighorn sheep living in North America in 1990 and 1991, the most recent years for which data has been published. Of this number, between twenty thousand and twenty-three thousand belong to the desert races, and the rest were mountain bighorn.

Another way of looking at the census is to examine the number of bighorn living in each country. Canada, from one estimate, has between fifteen and sixteen thousand individuals, while there are between forty-two thousand and forty-nine thousand bighorn in the United States. Mexico has about forty-five hundred of these animals. All of Mexico's sheep are desert bighorn and all of Canada's are mountain bighorn; only the United States has individuals of both types.

In the United States, these population estimates can be further broken down by state. Colorado has the most bighorn, with approximately 6,720 individuals. Five other states have over four thousand bighorn; in descending order they are Wyoming, Nevada, Idaho, Arizona, and California.

those additional large predators required prey. The result was an increased threat to endangered bighorn—and the loss of human life as well.

Because of this situation, in 1996 Californians were asked to vote on Proposition 197. If this referendum had passed, it would have amended the full protection granted mountain lions and provided a first step toward having a hunting season on the big cats. One of the state's largest newspapers, the *San Francisco Chronicle,* published an editorial urging voters to reject the proposition. In the weeks leading up to the vote, there was also a good deal of publicity about the human deaths attributed to mountain lions. Ultimately, Proposition 197 was rejected by California's voters.

Because of the controversy raised by the suggestion that mountain lions be culled—whether to prevent bighorn deaths or human fatalities—the Peninsular Bighorn Recovery plan is very careful about how it approaches this issue. The plan cautions: "Removal of mountain lions should be selective and only target individual lions known to be, or suspected of, preying on bighorn sheep."[15] Selective removal means not only that particular lions be killed but that such a policy would only be implemented in limited areas under specific criteria. For example, the plan recommends predator removal in a designated recovery region only if there are fewer than fifteen adult bighorn and if predation is a known mortality factor. Furthermore, since Proposition 197 was rejected, such removal could only be conducted by the California Department of Fish and Game. These policies ensure that killing mountain lions is a last resort and not a standard management tool.

The future of the bighorn

In nearly every portion of its range, the bighorn faces some measure of threat to its survival. Disease can destroy entire herds. Poaching remains a threat in some areas. And while there are a number of reserves providing sanctuary to bighorn, many bighorn herds exist in small numbers isolated from others of their kind. Small herds are particularly vulnerable to predation, inbreeding, and epizootic outbreaks.

Although bighorn face obstacles to their recovery, many individuals and organizations seek to minimize these obstacles. Translocations may be used to restore bighorn in some western parks, and captive breeding may help to augment herds of peninsular bighorn in southern California. Research continues and zoologists learn more about bighorn every year. Scientists, rangers, and teachers also give bighorn conservation a boost by educating the general public about current restoration projects. Success in these efforts might ensure that the bighorn persists as a distinctive resident of the mountains and deserts of western North America.

Notes

Chapter 1: Bighorn Natural History

1. John Hogg and Stephen Forbes, "Mating in Bighorn Sheep: Frequent Male Reproduction Via a High-Risk 'Unconvential' Tactic," *Behavioral Ecology and Sociobiology,* January 1997, p. 33.

2. Hogg and Forbes, "Mating in Bighorn Sheep," p. 33.

Chapter 2: Hunting and Poaching of Bighorn

3. Warren E. Kelly, "Hunting." Chap. 22 in Gale Monson and Lowell Sumner, eds., *The Desert Bighorn: Its Life History and Management.* Tucson: The University of Arizona Press, 1980, p. 337.

4. Lynn R. Irby, Jon E. Swenson, and Shawn T. Stewart, "Two Views of the Impacts of Poaching on Bighorn Sheep in the Upper Yellowstone Valley, Montana, USA," *Biological Conservation,* October 1989, p. 260.

Chapter 3: Human Encroachment Threatens the Bighorn

5. Francis J. Singer, Elizabeth Williams, Michael W. Miller, and Linda C. Zeigenfuss, "Population Growth, Fecundity, and Survivorship in Recovering Populations of Bighorn Sheep," *Restoration Ecology,* December 2000, p. 75.

6. William J. Foreyt, "Fatal *Pasteurella Haemolytica* Pneumonia in Bighorn Sheep After Direct Contact With Clinically Normal Domestic Sheep," *American Journal of Veterinary Research,* 1989, p. 341.

7. Thomas D. Bunch, Walter M. Boyce, Charles P. Hibler, William R. Lance, Terry R. Spraker, and Elizabeth S. Williams, "Diseases of North American Wild Sheep." Chap. 6 in Raul Valdez and Paul R. Krausman, *Mountain Sheep of North America.* Tucson: The University of Arizona Press, 1999, p. 216.

Chapter 4: Isolated Bighorn

8. Joel Berger, "Persistence of Different-Sized Populations: An Empirical Assessment of Rapid Extinctions in Bighorn Sheep," *Conservation Biology,* January 1990, p. 91.

9. John D. Wehausen, "Rapid Extinction of Mountain Sheep Populations Revisited," *Conservation Biology,* April 1999, p. 378.

Chapter 5: Bighorn Conservation

10. Francis J. Singer, Vernon C. Bleich, and Michelle A. Gudorf, "Restoration of Bighorn Sheep Metapopulations in and Near Western National Parks," *Restoration Ecology,* December 2000, p. 19.

11. Esther Rubin, "Recovery Plan for Bighorn Sheep in the Peninsular Ranges, California," *U.S. Fish and Wildlife Service,* 2000, p. 38.

12. Rubin, "Recovery Plan for Bighorn Sheep," p. 104.

13. Quoted in Mitch Tobin, "Lion Kill Would Aid Bighorns, State Says," *Arizona Daily Star,* July 18, 2001, p. A4.

14. Quoted in Tobin, "Lion Kill Would Aid Bighorns," p. A4.

15. Rubin, "Recovery Plan for Bighorn Sheep," p. 92.

Glossary

allogrooming: When one animal grooms another.

altitudinal migration: Movement between higher and lower elevations in response to seasonal changes.

aoudad: The barbary sheep of North Africa.

Artiodactyla: The order of mammals that includes even-toed hoofed mammals such as the bighorn.

atlatl: Spear-headed throwing sticks used by prehistoric humans to hunt bighorn and other game.

bachelor group: A group of male bighorn outside of breeding season.

Bergmann's Rule: The tendency for populations of a given species of warm-blooded animal to be larger in colder regions than in warmer regions.

blocking: One of three bighorn mating strategies. In blocking, a ram prods a ewe to move her away from the vicinity of other males.

bottleneck: The condition of having a small effective population size.

Bovidae: The family within order Artiodactyla that includes sheep, goats, cattle, and antelope.

broomed: The condition in which a bighorn breaks off only the tips of its horns.

browse: Leaves, shoots, and twigs of woody plants.

coursing: One of three bighorn mating strategies. A coursing male fights vigorously to gain the few seconds it needs to mate with a ewe.

critical: By the IUCN criteria, a species that has a 50 percent chance of going extinct within five years.

deleterious genes: Genes that can cause a potentially harmful condition in individuals possessing them.

delisting: When a species is removed from the U.S. Fish and Wildlife Service's endangered species list.

dominance hierarchy: An established order among social animals from the dominant individual to the least dominant individual.

downlisting: When a species is no longer designated as endangered by the U.S. Fish and Wildlife Service but is still listed under some other category, such as threatened.

ectoparasite: A parasite that lives on the outside of its host's body.

effective population size: In contrast to the actual population, this is a calculation of population based on how many animals of breeding age are in the group.

endangered: By the IUCN criteria, a species that has a 20 percent chance of going extinct within twenty years.

endoparasite: A parasite that lives within its host's body.

epizootic: An outbreak of disease among a population of animals.

escape terrain: Canyons, steep inclines, and other geological formations that afford the bighorn the best chance to escape from predators.

ewe group: See *nursery group;* ewe group is often used to describe subpopulations within the population of peninsular bighorn sheep in southern California.

forbs: Herbs and other low-growing broadleaf plants.

founders: The individuals present when a particular population originated.

fragmented: A distribution such as that of the bighorn; rather than having a continuous range, bighorn are found in isolated pockets of habitat.

genetic diversity: The degree of similarity or dissimilarity between individuals in a population. It is advantageous that genetic diversity be high.

geneticist: A scientist who studies genetics and the flow of genes in a population.

GIS: Geographic information system, any of several computer programs in which a site is mapped and areas that match specific criteria are located.

Gloger's Rule: Tendency in many animal species for individuals living in humid climates to be darkly colored and those in arid climates to be lightly colored.

herbivorous: An animal that eats vegetation.

home range: The area that an animal occupies as it conducts its daily activities.

inbreeding: Condition in which an animal mates with a close relative.

indigenous herd: A herd of bighorn that is native to the region it occupies.

IUCN: The World Conservation Union, an international organization that produces a list of endangered species.

keratin: Fiberous protein material that forms the outer part of the horns of a bighorn.

mutualism: A relationship between two organisms that benefits both.

necropsy: A veterinarian's examination of a dead animal to determine its cause of death.

nursery group: Nonbreeding season assemblage of ewes, lambs, and young rams.

parasitism: A relationship between two organisms in which one benefits and the other is negatively impacted.

pasteurellosis: Pneumonia caused by *Pasteurella* bacteria.

pathogens: Disease vectors such as bacteria and viruses.

Pleistocene: Period of the earth's history between 1.64 million years ago and ten thousand years ago. It included the Ice Ages.

prerut: Pronounced "pre-rut," that period before actual mating in which bighorn males have contests and fights to determine dominance.

radiotelemetry: Tracking an animal by signals emitted from a transmitter worn by the animal.

recovery plan: A detailed report issued by the U.S. Fish and Wildlife Service documenting the factors causing a species to be endangered and setting forth a process for bringing the population of the species up to a safe level.

rumen: The first chamber of the stomach of a ruminant.

ruminant: Artiodactyls, such as the bighorn, with a multi-chambered stomach and a complicated digestive process.

rut: The actual mating period in bighorn.

scabies: An infestation of mites in the ears.

selenium: A trace mineral needed by bighorn in small amounts for good growth.

sinusitis: An infection in the sinus caused by the nasal botfly.

taxonomist: A scientist who studies the classification and relationships among living things.

tending: One of three mating strategies in bighorn; a tending male stays close to the side of one or more females to keep other rams away from them.

translocation: To actively move bighorn to new areas, especially areas where they formerly lived but became extinct sometime in the recent past.

transplacental transmission: Where a lungworm larvae passes from a ewe into the fetus she is carrying.

transplant herd: A herd of bighorn composed of animals brought there from elsewhere by humans.

vulnerable: According to IUCN criteria, a species that has a 10 percent chance of going extinct within one hundred years.

Organizations to Contact

Bighorn Institute
PO Box 262, Palm Desert, CA 92261-0262
(760) 346-7334
www.bighorninstitute.org
An organization promoting the conservation of wild sheep. It supports research, particularly projects involving the peninsular bighorn sheep of southern California.

Foundation for North American Wild Sheep
720 Allen Avenue, Cody, WY 82414
(307) 527-6261
www.fnaws.org
This group is involved in the auction of bighorn hunting permits, with proceeds going to fund bighorn conservation programs.

National Bighorn Sheep Center
PO Box 1435, Dubois, WY 82513
(888) 209-2795
www.bighorn.org
This organization is dedicated to educating the public about the habitat and conservation needs of Rocky Mountain bighorn.

Society for the Conservation of Bighorn Sheep
PO Box 94182, Pasadena, CA 91109-4182
(323) 256-0463
www.desertbighorn.cjb.net
An organization dedicated to the conservation and management of California's desert bighorn sheep.

For Further Reading

Books

Roger Few, *Animal Watch.* New York: DK Publishing, 2001. A well-illustrated book about protecting endangered species and their habitats.

John Muir, *The Mountains of California.* New York: The Modern Library, 2001. Muir was a prominent nineteenth-century naturalist. First published in 1894, chapter fourteen of the book is entitled "Wild Sheep." Here, Muir shares details of his encounters with bighorn in the Sierra Nevada.

Dorothy Hinshaw Patent, *Back to the Wild.* San Diego: Gulliver Books, 1997. Conservation programs that involve breeding specimens in captivity for release into the wild are the focus of this volume.

Russell Roberts, *Endangered Species.* San Diego: Lucent Books, 1998. A generalized, issue-based look at conservation of endangered species.

Victoria Sherrow, *Endangered Mammals of North America* (Scientific American Sourcebooks). Brookfield, CT: Twenty First Century Books, 1995. Black-footed ferrets, Florida panthers, red wolves, and other declining North American mammals are examined in this book.

J. David Taylor, *Endangered Desert Animals.* New York: Crabtree Publishing, 1992. A number of endangered animal species from around the world are discussed and illustrated in this book. Includes a chapter on the desert bighorn.

J. David Taylor, *Endangered Mountain Animals.* New York: Crabtree Publishing, 1992. A number of endangered mountain animals from around the world are discussed, including the mountain bighorn.

Periodicals

Marla Cone and Diana Marcum, "Desert Bighorns to Get U.S. Protection," *Los Angeles Times,* March 13, 1998, p. A3. The reporters interviewed developers, elected officials, and conservationists to get their reactions to the U.S. Fish and Wildlife Service's decision to add the peninsular bighorn to the federal government's list of endangered species.

Veronique de Turenne, "Bighorns Being Fenced Out to Save Their Lives," *Los Angeles Times,* January 2, 2002, pt. 2 p. 6. Describes a $1.2 million project to build a three-and-a-half-mile long fence at Rancho Mirage, California, to keep peninsular bighorn from straying into urban areas.

Jim Erickson, "Radio Collars to Keep Track of Bighorn Sheep Study to Provide Best Count Yet in National Park," *Rocky Mountain News,* August 11, 2001, p. 16A. This article descibes a Colorado State University project to use radio collars on sixty bighorn in Rocky Mountain National Park in Colorado.

Bob Holmes, "Up Against Steep Odds," *National Wildlife,* February 1997, p. 46. A look at the plight of the peninsular bighorn of southern California.

David N.B. Lee, "Back Where They Belong: Reintroduction of Bighorn Sheep Into Parks," *National Parks,* September 1998, p. 26. A look at bighorn translocations.

Thomas A. Lewis and Jeff Vanuga, "The Town That Cried Sheep," *National Wildlife,* February 1995, p. 8. The story of how the town of Dubois, Wyoming, boosted its local economy by promoting bighorn-related ecotourism.

Diana Marcum, "Sheep, Golfers Butt Heads Over Use of Desert," *Los Angeles Times,* March 16, 1997, p. A3. This article covers the controversy over construction of golf courses in peninsular bighorn habitat in southern California.

Seattle Times, "Oregon Short of Goal for Moving Bighorns," February 9, 2001, p. B5. Describes an ongoing bighorn translocation project in the Pacific Northwest.

Works Consulted

Books

Paul Cutright, *Lewis and Clark: Pioneering Naturalists.* Urbana: University of Illinois Press, 1969. A description of the famous expedition to the Pacific Northwest, with excerpts from Lewis and Clark's own journals. Pages 133 to 134 contain a brief account of the expedition's first encounter with bighorn.

Jared Diamond, *The Third Chimpanzee.* New York: Harper Collins, 1992. Chapter eighteen discusses the "blitzkrieg" theory—the idea that aboriginal humans caused mass extinctions of large North America mammals in the Pleistocene age.

Alan Feduccia, *The Origin and Evolution of Birds, 2nd ed.* New Haven: Yale University Press, 1999. Although primarily a book about birds, pages 309 to 311 offer a concise look at the Pleistocene extinctions that affected mammals as well as birds.

V. Feuerstein, et al., *Bighorn Sheep Mortality in the Taylor River-Almont Triangle Area 1978–1979: A Case Study.* Colorado Division of Wildlife Special Report, 1980. An examination of the effects of disease on bighorn survival.

Valerius Geist, *Mountain Sheep: A Study in Behavior and Evolution.* Chicago: The University of Chicago Press, 1971. The first extensive book to be published about bighorn sheep.

Gale Monson and Lowell Sumner, eds. *The Desert Bighorn: Its Life History and Management.* Tucson: University of Arizona Press, 1980. Describes in detail all aspects of the natural history of bighorn native to the arid portions of the southwestern United States and Mexico.

Ronald Nowak and Ernest Walker, *Walker's Mammals of the World, 6th ed.* Baltimore: Johns Hopkins University Press,

1999. This is a detailed encyclopedia of mammals, published in two volumes. Pages 1231 to 1238 of Volume 2 discuss bighorn and other members of the genus *Ovis.*

Esther Rubin, *Recovery Plan for Bighorn Sheep in the Peninsular Ranges, California.* Portland, OR: U.S. Fish and Wildlife Service, 2000. This document lists threats to the bighorn of southern California and proposes methods of stabilizing and increasing their population.

David Shackleton, ed., *Wild Sheep and Goats and Their Relatives: Status Survey and Conservation Action Plan for Caprinae.* Switzerland: IUCN, 1997. A publication describing the threats to various species of wild goats and sheep worldwide. It is subdivided not by species but by countries, so bighorn conservation is discussed separately in the sections on Canada, Mexico, and the United States.

C.A. Spinage, *The Natural History of Antelopes.* New York: Facts On File, 1986. While this book is primarily about the ecology of African antelope, pages 19 to 22 provide a good introduction to the digestive process of all ruminants.

Raul Valdez and Paul Krausman, eds., *Mountain Sheep of North America.* Tucson: University of Arizona Press, 1999. This recently published book has up-to-date information on all aspects of the biology and management of bighorn as well as their northern cousins, the thinhorn.

Periodicals

Joel Berger, "Persistence of Different-Sized Populations: An Empirical Assessment of Rapid Extinctions in Bighorn Sheep," *Conservation Biology,* March 1990.

Gary Brundige, "Fatal Fall by Bighorn Lamb," *Journal of Mammalogy,* May 1987.

Ann S. Causey, "On the Morality of Hunting," *Environmental Ethics,* Winter 1989.

J.C. Driver, "Early Prehistoric Killing of Bighorn Sheep in the Southeastern Canadian Rockies," *Plains Anthropologist,* April 1982.

William J. Foreyt, "Reproduction of Rocky Mountain Bighorn Sheep in Washington: Birth Dates, Yearling Ram Reproduction and Neonatal Diseases," *Northwest Science,* Fall 1988.

————, "Fatal *Pasteurella Haemolytica* Pneumonia in Bighorn Sheep After Direct Contact With Clinically Normal Domestic Sheep," *American Journal of Veterinary Research,* March 1989.

Martin Forstenzer, "Bighorn Sheep Losing Ground, and Lives, to an Old Foe," *New York Times,* September 29, 1998.

D. Ganskopp and M. Vavra, "Slope Use by Cattle, Feral Horses, Deer, and Bighorn Sheep," *Northwest Science,* April 1987.

Christine C. Hass, "Bighorn Lamb Mortality: Predation, Inbreeding, and Population Effects," *Canadian Journal of Zoology,* March 1989.

————, "Alternative Maternal Care Patterns in Two Herds of Bighorn Sheep," *Journal of Mammalogy,* January 1990.

————, "Seasonality of Births in Bighorn Sheep," *Journal of Mammalogy,* November 1997.

Christine C. Hass and Donald Jenni, "Social Play Among Juvenile Bighorn Sheep: Structure, Development, and Relationship to Adult Behavior," *Ethology,* April 1993.

Charles L. Hayes, et al., "Mountain Lion Predation of Bighorn Sheep in the Peninsular Ranges, California," *Journal of Wildlife Management,* October 2000.

Thomas A. Heberlein, "Changing Attitudes and Funding for Wildlife: Preserving the Sport Hunter," *Wildlife Society Bulletin,* Winter 1991.

P.W. Hedrick, et al., "Founder Effect in an Island Population of Bighorn Sheep," *Molecular Ecology,* April 2001.

Loren L. Hicks and James M. Elder, "Human Disturbance of Sierra Nevada Bighorn Sheep," *Journal of Wildlife Management,* October 1979.

Manfred Hoefs and Uli Nowlan, "Hybridization of Thinhorn and Bighorn Sheep," *Canadian Field Naturalist,* October 1997.

John Hogg, "Mating in Bighorn Sheep: Multiple Creative Male Strategies," *Science,* August 3, 1984.

John Hogg and Stephen Forbes, "Mating in Bighorn Sheep: Frequent Male Reproduction Via a High-Risk 'Unconventional' Tactic," *Behavioral Ecology and Sociobiology,* January 1997.

David J. Huggard, "Prey Selectivity of Wolves in Banff National Park: I. Prey Species," *Canadian Journal of Zoology,* January 1993.

Lynn R. Irby, et al., "Two Views of the Impacts of Poaching on Bighorn Sheep in the Upper Yellowstone Valley, Montana, USA," *Biological Conservation,* April 1989.

Jon T. Jorgenson, et al., "Harvesting Bighorn Ewes: Consequences for Population Size and Trophy Ram Production," *Journal of Wildlife Management,* July 1993.

————, "Effects of Age, Sex, Disease, and Density on Survival of Bighorn Sheep," *Ecology,* June 1997.

————, "Effects of Population Density on Horn Development in Bighorn Rams," *Journal of Wildlife Management,* July 1998.

K.A. Keating, "Allogrooming by Rocky Mountain Bighorn Sheep, Glacier National Park, Montana," *Canadian Field Naturalist,* January 1994.

M.D. Kock, et al., "Capture Methods in Five Subspecies of Free-Ranging Bighorn Sheep: An Evaluation of Drop-Net, Drive-Net, Chemical Immobilization and the Net-Gun," *Journal of Wildlife Diseases,* Fall 1987.

B.D. Leopold and Paul R. Krausman, "Status of Bighorn Sheep in Texas," *Texas Journal of Science,* no. 2, 1983.

Gary D. Miller and William S. Gaud, "Composition and Variability of Desert Bighorn Sheep Diets," *Journal of Wildlife Management,* July 1989.

Michael W. Miller, et al., "Drug Treatment for Lungworm in Bighorn Sheep: Reevaluation of a 20-Year-Old Management Prescription," *Journal of Wildlife Management,* April 2000.

Fen Montaigne, "Paying Big Bucks for a Shot at a Bighorn," *Wall Street Journal,* January 9, 1998.

Linda W. Norrix, et al., "Conductive Hearing Loss in a Bighorn Sheep," *Journal of Wildlife Diseases,* April 1995.

Loreann S.A. Pendleton and David Hurst Thomas, "The Fort Sage Drift Fence, Washoe County, Nevada," *Anthropological Papers of the American Museum of Natural History,* vol. 58, pt. 2, 1983.

Marc Peyser, "Predators on the Prowl; Bloody Cougar Attacks Trouble the Booming West," *Newsweek,* January 8, 1996.

Gary Polakovic, "Fighting a Battle for Little Bighorns," *Los Angeles Times,* August 29, 2001.

Rob Roy Ramey II, et al., "Genetic Bottlenecks Resulting from Restoration Efforts: The Case of Bighorn Sheep in Badlands National Park," *Restoration Ecology,* December 2000.

P. Ian Ross, et al., "Cougar Predation on Bighorn Sheep in Southwestern Alberta During Winter," *Canadian Journal of Zoology,* May 1997.

Esther Rubin, et al., "Reproductive Strategies of Desert Bighorn Sheep," *Journal of Mammalogy,* August 2000.

Kathreen Ruckstuhl, "Foraging Behaviour and Sexual Segregation in Bighorn Sheep," *Animal Behaviour,* July 1998.

San Francisco Chronicle, "Protect the Cougars—Vote No on Prop. 197," February 16, 1996.

David Shackleton and J. Haywood, "Early Mother-Young Interactions in California Bighorn Sheep," *Canadian Journal of Zoology,* April 1985.

Francis J. Singer, et al., "Restoration of Bighorn Sheep Metapopulations in and Near Western National Parks," *Restoration Ecology,* December 2000.

———, "Population Growth, Fecundity, and Survivorship in Recovering Populations of Bighorn Sheep," *Restoration Ecology,* December 2000.

———, "Translocations as a Tool for Restoring Populations of Bighorn Sheep," *Restoration Ecology,* December 2000.

Craig A. Stockwell, "Behavioural Reactions of Desert Bighorn Sheep to Avian Scavengers," *Journal of Zoology,* December 1991.

Craig A. Stockwell, et al., "Conflicts in National Parks: A Case Study of Helicopters and Bighorn Sheep Time Budgets at the Grand Canyon," *Biological Conservation,* June 1991.

Mitch Tobin, "Lion Kill Would Aid Bighorns, State Says," *Arizona Daily Star,* July 18, 2001.

J.C. Turner, "Osmotic Fragility of Desert Bighorn Sheep Red Blood Cells," *Comparative Biochemistry and Physiology,* January 1979.

A.C.S. Ward, et al., "Immunologic Responses of Domestic and Bighorn Sheep to a Multivalent *Pasteurella haemolytica* Vaccine," *Journal of Wildlife Diseases,* April 1999.

Greg D. Warrick and Paul R. Krausman, "Barrel Cacti Consumption by Desert Bighorn Sheep," *Southwestern Naturalist,* October 1989.

John D. Wehausen, "Rapid Extinction of Mountain Sheep Populations Revisited," *Conservation Biology,* April 1999.

John D. Wehausen and Rob Roy Ramey II, "Cranial Morphometric and Evolutionary Relationships in the Northern Range of *Ovis Canadensis,*" *Journal of Mammalogy,* February 2000.

G.W. Welsh and T.D. Bunch, "Psoroptic Scabies in Desert Bighorn Sheep from Northwestern Arizona," *Journal of Wildlife Diseases,* Fall 1983.

Brian Wilkeem and Michael Pitt, "Diet of California Bighorn Sheep in British Columbia: Assessing Optimal Foraging Habitat," *Canadian Field Naturalist,* July 1992.

Index

Picture Credits

About the Author

Brett Bannor has been a zookeeper for twenty years and currently works at Zoo Atlanta. He has written several papers that have appeared in zoological journals and is one of the authors of the life history of the Common Moorhen, part of the *Birds of North America* series. Brett lives in Atlanta, Georgia, and enjoys reading about law, architecture, and history as well as zoology.